Customer Experience Management for Water Utilities

Customer Experience Management for Water Utilities

Marketing Urban Water Supply

Peter Prevos

IWA PUBLISHING

Published by **IWA Publishing**
Alliance House
12 Caxton Street
London SW1H 0QS, UK
Telephone: +44 (0)20 7654 5500
Fax: +44 (0)20 7654 5555
Email: publications@iwap.co.uk
Web: www.iwapublishing.com

First published 2018

British Library Cataloguing in Publication Data
A CIP catalogue record for this book is available from the British Library

ISBN: 9781780408668 (Paperback)
ISBN: 9781780408675 (eBook)
ISBN: 9781780408897 (*ePub*)

Cover image: www.istockphoto.com

Contents

List of Figures

List of Tables

About the Author

Peter Prevos is a civil engineer and social scientist. He obtained a bachelor of civil and environmental engineering from the Zuyd University of Applied Sciences in Heerlen, the Netherlands. During the first phase of his career he worked as a structural engineer in Johannesburg and as a project engineer for the Royal Boskalis Westminster dredging company. With Boskalis, Peter worked on land reclamation and flood mitigation projects in Europe and South-East Asia.

After returning to the Netherlands, Peter worked for Rijkswaterstaat, the Dutch government agency responsible for the design, construction, and operation of major road and water infrastructure facilities. For this organisation, Peter was a senior project manager at the Maaswerken project, a flood mitigation project in the Southern Netherlands.

Since 2000, Peter lives in Australia where he works for Coliban Water in regional Victoria. For this water utility, he managed water and sewer infrastructure projects, implemented integrated business improvement systems and managed the land development function. During that time, Peter obtained a Bachelor of Arts from Monash University and an MBA from La Trobe University. Peter is currently responsible for the data science function, where he and his team create value from data about water systems, assets and customers.

His work with the land development process introduced him to the science of marketing as a tool to implement customer-centric processes. For five years Peter has researched customer centricity in water utilities and published the results in a PhD dissertation named *The Invisible Water Utility: Employee Behaviour and Customer Experience in Service-Dominant Logic.* This dissertation discusses the applicability of marketing theory to water utilities.

Preface

This book is the culmination of a career in engineering in which I slowly moved from the technical to appreciating the social dimension. After graduation I worked for a multinational marine engineering contractor, reclaiming land in Europe and Asia. That period of my career was largely a mathematical endeavour, motivated by maximising the efficiency and effectiveness of construction processes. The needs and wants of the people who would eventually use the structures I helped to build were not front-of-mind.

When returning to the Netherlands, I joined a government agency responsible for national water infrastructure. The transition from a hyper-competitive and profit motivated industry to a public service organisation was in some ways a bigger culture shock than I had previously experienced working outside Europe. My experiences with vastly different national and organisational cultures, combined with an academic study of the humanities, sparked my interest in management as a science.

After completing a Bachelor of Arts, it was time to deepen my knowledge of management by undertaking an MBA. The marketing lectures by Professor Rhett Walker were a career-changing experience. Before taking this course, my views on marketing were largely shaped by the neo-Marxist philosophy of the Frankfurt School, which perceives marketing as a source of many societal problems. Professor Walker's lectures convinced me that marketing encompasses more than selling for the sake of profits and creating false needs. I came to recognise marketing as a tool to maximise the value that organisations contribute to society.

While undertaking an MBA, I was also responsible for managing a team providing services to property developers. When I first started this task, relationships with developers were not very strong as we positioned ourselves as a regulatory authority instead of providing a service. Implementing the principles of marketing enabled me to improve the level of service given to the developers of land. This experience caused a Copernican Revolution in my understanding on what it means to be a civil engineer, moving towards integrating the humanities with the physical sciences. This broader thinking about marketing enabled me to combine my interest in the humanities with my profession. Combining the procedural thinking of engineering with the philosophical method of enquiry is a powerful pathway to foster innovation.

Undertaking research in customer service and investigating the dynamics of human interaction doesn't seem to match the social identity of an engineer. As an engineer I am wrapped in stereotypes of socially inept professionals, highly focused on the technical task at hand, but lacking people skills. For my engineering degree, I took a compulsory subject on organisational behaviour. It was the least popular subject in the engineering curriculum as it was perceived to be 'fuzzy' and lacking the certainty common to other subjects.

A civil engineer designs and builds for people, the *civilis*, and not for the sake of engineering itself. The value of engineering lies not in its tangible results but in the intangible benefits that society derives from it. The importance of the intangible benefits of tap water is highlighted by the realisation that many inspirational moments during this research came to me when standing in the shower.

The choice to continue towards a PhD on that very topic was born in a desire to deepen my understanding of the social science foundations of my profession. My motivation was also based on the observation that engineering decisions in water utilities are guided by theory, while decisions on matters pertaining to customer relationships are devoid of such foundations and based on experience and 'common sense'. This book seeks to develop a framework for water utility managers to guide their decisions in the realm of the social sciences.

This book is guided by my experiences as a manager of various functions in a water utility and by my academic research in this area. I sought a balance between managerial relevance and theoretical sophistication. Being both an engineer and social scientist, this book is driven by a mixed methodology, following subjectivism as its ontological foundation. Although engineering is dominated by positivist notions of predictability and is very successful at using this method, marketing is a social science where subjectivity is inescapable. My personal objective in writing this book is to bridge the divide between the physical sciences that dominate our industry and the social sciences that are needed to become more focused on customers.

The text in this book is largely the same as my dissertation on this topic. The main difference between the two works is that the dissertation is written for marketing academics, while this book is written for water utility managers. Detailed discussions about methodology and marketing theory have been replaced with practical examples and reflections. Readers who are interested in the theoretical and methodological background of this research are encouraged to read the dissertation.

Writing this book has been a great adventure that allowed me to combine my professional experience as a water engineer with theoretical insights in the science and practice of marketing, inspired by the humanities. As an engineer and water professional I am interested in how my profession influences society and how the theories of marketing can be applied to this industry to maximise value for the community. As a social scientist, I am interested in the role of marketing in contemporary society, with specific reference to public service. The combination of engineering and marketing are not as incompatible as it seems. Marketing is after all, in the words of Kotler and Levy (1969): 'customer satisfaction engineering'.

Foreword

Melanie Goetz, Hughes & Stuart Communications

The time has come for tap water to flow out of the shadows of darkness and into the sunlight of customer appreciation. In an invigorating flood of insight, Peter Prevos argues that the complexities of water management boil down a series of simple paradoxes that essentially teach us a simple lesson: People know water is essential to survival—they rage against even the smallest rate hike and rail about perceived water quality shortcomings—yet remain utterly oblivious to our entire industry on a day-to-day basis.

Utility managers: If you really want to engage with the public and humanise the vital search for water resources and the genius of engineering for the long-term use and preservation of those resources, turn now to the opening page of this book, it is a must read.

If you seek fresh understanding of the intricate interplay between your life's work and the lives of millions of water consumers, carve out the time to read and absorb these pages. Don't be fooled by Peter's long list of academic credits and achievements; don't let his engineering background or technical expertise intimidate you. In these pages, Peter offers countless revelations about a topic that is at once so simple, yet upon reflection so delightfully intricate and complex—the public's perception of water quality.

Before writing this book, Peter researched customer centricity in water utilities as a PhD dissertation. This book enriches his dissertation theories with the experiences of a career in engineering, working with marketers, managers, citizens, and politicians. And yes, though Peter spent his early career submersed in a world of mathematics, he'll never drown you in marketing theory, formulas and hypotheses.

In a delicate balancing act, he actually has created a book—a masterpiece—that offers practical insights in making your water utility visible, relevant, and (perhaps most importantly) understood.

Peter knows well that gaining strong customer approval requires more than a government-stamped declaration that water is 'safe to drink' … it must also be clear and refreshing. It can't be discoloured. It can't stink. And above all, it has to taste good.

I first encountered Peter Prevos when he published a review of my first book, *Communicating Water's Value: Talking Points, Tips & Strategies*, (Goetz, 2014a). I was certainly flattered, but also curious about the things he pointed out about 'the customer's experience.' A while later, I heard Peter was initiating a 'Tap Crawl', a tongue-in-cheek name for a program to promote the importance of the taste of water. A Tap Crawl is designed to work both internally, with a utility's employees and stakeholders, and externally with the general public.

What do utility insiders think about the taste, appearance and purity of their water? What does the average person think about the water coming out of their kitchen faucet?

I was intrigued, and I wanted to learn more. One thing led to another, and we ended up collaborating on 'Tap Crawl from Down Under', an article in the December 2016 issue of the *Journal of the American Water Works Association* (Goetz & Prevos, 2016). Hopefully our work will invite others to investigate the multi-layered relationship between supplier and customer and enrich the partners' sharing of our most precious global resource.

Admittedly, it's tough to deal with the myriad factual details and perceptual variables involved in acquiring, processing, distributing and recycling our water resources. But in the long run, enlightening people about water quality will pay off; shining a spotlight on the water system's now largely invisible contribution to the public health and welfare. If you manage, govern, operate, or handle public outreach for a utility, (or have more than a passing interest in such worthy topics), I encourage you to read on. This book is a treasure.

About Melanie K. Goetz

International speaker, consultant, and author of *Communicating Water's Value: Talking Points, Tips & Strategies* (2014) and *Communicating Water's Value Part 2: Waste water, Stormwater & Watersheds* (2017).

Melanie specialises in helping water-related entities to effectively communicate with their customers, elected officials, the media, and stakeholders. President of Hughes & Stuart Communications, she holds an MBA from Old Dominion University, Virginia, USA.

Acknowledgements

The countless hours of solitary intellectual labour notwithstanding, this book and its associated dissertation could not have been realised without the help of many people. Several important concepts developed herein are based on the wisdom of fellow water professionals and although they are not formally cited, these contributions nevertheless deserve recognition.

My first inspiration to think about what constitutes customer service in water utilities comes from Geoff Michell, former Managing Director of Coliban Water. During the Millennium Drought, the organisation was under a lot of pressure from the community. Geoff once told me that he wished that things were they way the used to be, a time when nobody cared about water. This insight inspired me to introduce the concept of the *Invisible Water Utility*. This idea was not further developed until some years later when I took the marketing lectures of Emeritus Professor Rhett Walker. He convinced me that the study of marketing is a worthwhile intellectual pursuit. Together with former colleague and study partner Ian Watson, we took the first steps in this topic through our joint MBA projects on water utility marketing. Ian's creative thinking greatly influenced some of the concepts in this book.

My former manager Dr Dharma Dharmabalan introduced me to the difference between 'Good Water' and 'Safe Water', which allowed me to develop the Service Quality concepts. Jeff Rigby, current Managing Director of Coliban Water, often says that regulators should be viewed as customers that benefit from being provided with information. His thinking has inspired me to include the concept of value creation networks and recognise the importance of Balanced Centricity, which moves beyond the traditional dyadic customer relationships.

Throughout this intellectual journey, I have also been supported by professional networks. Tony Wright of VicWater in Melbourne invited me twice to speak at their annual conferences to share my progress with the wider industry, which helped me to focus my research. The Australian Water Association and the International

Water Association have been helpful in providing contacts to discuss issues or to recruit respondents for this research. My current and past managers Jeff Rigby, David Sheehan, Dr Dharma Dharmabalan and Gavin Hanlon have provided me with generous opportunities to undertake this herculean task. My academic supervisors Prof. Gillian Sullivan Mort and Dr Clare D'Souza and past supervisors Dr Navin Veerapa and Prof. Richard Tay have enabled me to shape my dissertation and provided me with intellectual freedom and independence to pursue a non-traditional approach.

Social media has also played an important role in developing these ideas. My connections on the LinkedIn social network, followers of the @*InvisibleH2O* Twitter feed and readers of the *invisiblewater.org* blog helped me fine-tune these ideas.

These acknowledgements would not be complete without recognising Mark Hammond and the team from IWA publishing for agreeing to publish a book about the social science of water. Melanie Goetz, Ian Monks, Clare Lombardi, David Sheehan and Mick Dunne who were so kind to critically read some of the chapters.

Last, but not least I would like to acknowledge the hundreds of anonymous employees and customers of water utilities who donated their time to provide me with the information that forms the foundation of this research.

Chapter 1

Introduction to water utility marketing

Water is one of the most essential, but often misunderstood, natural resources. The spectre of water shortages in many areas around the world has increased the attention for water at an international level (Fagan, 2011; Zetland, 2011). This attention is largely aimed at the future uncertainty of water supplies due to population growth and climate change. Whilst the availability of water is an urgent problem in need of solutions, regulators and society urge water utilities to become more focused on customers. This focus on customers goes beyond providing a reliable water supply through technological solutions. Recent societal developments have raised the expectations that customers have of all service providers, including water utilities.

Water utilities have repeatedly been criticised in economics and business literature for lacking customer focus (Ahmed, 2009; Auriol & Picard, 2009; Deichmann & Lall, 2007; Skellett, 1995). There is also societal pressure on water utilities to become more focused on the needs of their customers. Customer advocacy groups lobby the industry and government to ensure that water utilities improve the quality of their engagement with customers. Also, government regulators increasingly require detailed performance reports, including metrics on customer service (Ben-David, 2015; CUAC, 2013; Essential Services Commission, 2012; Falp & Le Masurier, 2009; Hall Partners & Open Mind, 2011).

This increased attention from regulators on how utilities relate to customers has changed the industry profoundly. Regulatory frameworks often require water utilities to initiate customer service programmes. These programmes, however, are often more motivated by the reporting requirements of the regulator rather

Peter Prevos
doi: 10.2166/9781780408675_001

than by imperatives emerging from within the utilities (Karbowiak, 2003). Economic regulation acts as a proxy for the intrinsic motivational mechanisms that are present in competitive markets to drive service providers to become more focused on maximising value for their customers. Regulators can motivate or enforce behaviours in water utilities that mimic the behaviour in competitive markets.

The criticism of lacking customer focus notwithstanding, water utilities are some of the most reliable service providers. Taxi company Uber is often lauded as an example of a market disrupter through their improvement of their industry's value proposition. It is interesting to note that this successful company is in fact inspired by the high level of service provided by water utilities. The Uber website states that they aspire to become 'just as reliable as running water' (Ferenstein, 2015).

Tap water is provided directly into the houses of consumers and the service is so reliable that it is taken for granted and has become a symbol of a modern civilised life (Allon, 2006). In places with well-functioning water supplies, water services are normalised and reside in the background of daily life. Tap water is so reliable that customers don't need to spare any thought about purchasing and consuming it. Consumers pay a very low time-price for their water, which renders it invisible to their daily lives. The time-price is the amount of time that customers use to enjoy their water service. This time-price includes time spent on billing, disruptions, complaints and so on. In tap water, the time-price is close to zero because customers only need to open their tap. In areas without improved water service, this time-price can amount to several hours per day. Given that value is a function of the total cost to a customer, the lower the time-price, the higher the level of service.

The dissertation on which this book is based develops the notion of the *Invisible Water Utility* (Prevos, 2016). This idea refers to the fact that the providers of tap water are invisible to consumers. Services provided by these organisations are neither differentiated nor branded and are enjoyed as a matter of course. Although this normalisation of water is one of the great achievements of the past century, it has also disconnected users from the source of water and from the organisations that provide this precious resource to their homes.

The invisibility of the process of service provision, combined with the fact that water utilities are monopolies has also rendered them invisible to business scholars, including the field of marketing (Kurland & Zell, 2010; Patsiaouras *et al.*, 2015). The central theme of this book is that the invisibility of this industry in marketing scholarship and practice needs rectification.

The managerial objective of this book is to enhance the current discourse on service-centric provision of tap water by proposing a framework based on marketing theory. This framework has been developed specifically for this sector to assist managers with maximising the value they provide to customers. This book advocates that using marketing theory is of value in the same way that physics and biology is used to provide core services. It is not argued that social sciences should replace the physical sciences, or that they are superior. The social sciences and the physical sciences are both necessary to manage a successful water utility.

The word theory often has negative connotations. The phrase 'It's only a theory' is often used to indicate that deep thinking has little value to the reality of business. Managers seek practical ideas by mimicking what is considered best practice. Following what is ostensibly best practice only results in what is best within the current context. To improve beyond what is current best practice, businesses need to go beyond what is commonly known. Generating new best practice can be achieved by developing a framework based on 'first principles'. The framework presented in this book is based on marketing theory which describes the essence of delivering value.

The motto for this book is that 'nothing is so practical as a good theory', which was coined by Kurt Lewin (1951), a pioneer in organisational psychology. The essence of this statement is that a good grasp of theory, either sociology, psychology, biology or physics, assists managers to maximise the value of their decisions.

The framework presented in this book is based on Service-Dominant Logic (S-D Logic), a contemporary theory for the marketing of goods and services (Vargo & Lusch, 2004, 2008, 2016). This model for marketing emphasises that value is co-created between the service provider and the customer and that goods are mechanisms to provide intangible services.

S-D Logic has attracted a considerable amount of attention in marketing literature, but most contributions have been developed at a theoretical level. The dissertation on which this book is based enhances S-D Logic theory to render it more suitable for practical implementation, using water utilities as a case study (Prevos, 2016). This first chapter introduces marketing from a theoretical perspective, focusing on S-D Logic and illustrates how this relates to the management of water utilities.

This book is firmly grounded in managerial practice and seeks to bridge the gap between the natural science dominated discourse in water utilities and social science based marketing theories. The distance between the two disciplines might not be as large as it would seem. Marketing is after all, in the words of Kotler and Levy (1969) 'customer satisfaction engineering'.

1.1 MARKETING THEORY AND PRACTICE

Analysing tap water from a marketing perspective might seem trivial because there is no competitive substitute product. However, detailed knowledge and understanding of the role of water in contemporary society can assist utility managers in improving the value provided to their customers. Contemporary theories of marketing are no longer focused on techniques to maximise sales, but on developing processes that maximise value as perceived by the customer. Contemporary marketing applies just as much to the public sector and non-profit organisations as it does to the traditional commercial markets.

The application of marketing theory within water utilities tends to be sparse as marketing is viewed to be irrelevant for monopolistic service providers and public services (Brown, 2010b; Laing, 2003). This perception prevails because practitioners tend to view marketing from its narrow definition of promotion and selling, rather than the broader context of creating value for customers (American Marketing Association, 2013; Harrison & Stamp, 1991). The uneasy status of marketing in water utilities was hinted at in episode 13 of *The Gruen Transfer* (2009), an Australian television show about marketing and society. When discussing why bottled water is much more expensive than tap water, panel member Russel Howcroft exclaimed that water utilities are 'lazy marketers'. This hyperbolic accusation contains a kernel of truth as water utilities usual focus on technology and marketing is considered peripheral (Bell, 2012; Brown, 2010b). This book proposes a framework to overcome Howcroft's critique by embedding marketing theory into the strategic direction of water utilities.

The core competencies of water utilities are grounded in science, technology, engineering, and mathematics (STEM). The provision of clean and reliable water services is a complex process that relies on physics and biology as a theoretical framework to choose the best course of action. The paradigms of these sciences are based on objective analysis of physical facts, which can lead to the social dimension of water being suppressed. The social sciences are the theoretical foundations of marketing, which acknowledge the subjectivity of the human experience.

Although applying the physical sciences is necessary to ensure that water utilities are able to deliver the high level of reliability praised by Uber, the STEM approach alone is not sufficient to maximise customer value. The social sciences are able to provide a human context to the activities of water utilities that a reliance on technology alone is unable to achieve.

1.1.1 Definitions of marketing

When most people use the word marketing they are referring to selling and advertising. Contemporary understanding of marketing is, however, much broader than this limited point of view. Marketing is the process of creating value and within the context of water utilities, this value flows from the catchments to the customer's tap, their bills and beyond.

The term marketing is often also associated with negative connotations such as deceptive advertising, pressure selling and other unethical practises (Wickham, 2009). Marketing is a driving force behind consumerism and is often blamed for many of the ills of the modern world. This criticism aside, the field of marketing has, over the past few decades, developed into a discipline that is much broader than simply maximising the sales of good and services. The evolution from a sales-driven discipline to the study of how value is created is expressed in the various definitions of marketing. Early definitions of marketing describe it as the process of maximising sales while later versions shift the attention to maximising customer satisfaction. The most recent definition of marketing published by the American Marketing Association (2013) moves to a broader scope and defines the discipline as:

> ...the activity, set of institutions, and processes for creating, communicating, delivering, and exchanging offerings that have value for customers, clients, partners, and society at large.

This definition is based on the societal marketing concept which places the organisation in a symbiotic relationship with its customers and community overall. This definition emphasises that marketing is a process where one party creates value for another, moving beyond simple dyadic customer-firm relationships. The idea that marketing is not only suitable to maximise sales but that it can also be applied to societal issues was first proposed by Kotler and Levy (1969). They positioned marketing as a form of social engineering which involves influencing social parameters to satisfy the needs of society as a whole.

Contemporary views of marketing have lessened the importance of customer satisfaction and moved their attention to maximising value. In this holistic view, marketing is not an activity undertaken by a specialised department but an approach that applies to the whole of the organisation. Management guru Peter Drucker (1954, emphasis added) said it best when he wrote:

> Marketing is so basic that it cannot be considered to be a separate function. It is the *whole business* seen from the point of view of its final result, that is, from the *customer's point of view*.

1.1.2 Service-dominant logic

Just like engineering and science are guided by the 'laws of nature', marketing is based on the social sciences. The difference between these two ways of thinking is that engineering and science use mathematical models to *explain* and predict the physical reality, while the social sciences use conceptual ideas to *understand* society.

S-D Logic is a theoretical model of the exchange of value that marketers can use to better understand the market in which they operate. This concept is founded on the idea that customers don't purchase goods or services, but they buy services that create value. A fundamental principle of consumer behaviour is that customers don't purchase things for what they do, but for what they mean to them. For example, a three-piece suit is not just a set of clothes to protect your body from the heat or cold, it also provides the service of establishing your public self. The suit not only keeps you warm, it also communicates status and trust. Value is not solely created in the exchange of a product or a service but is created in the use of these. Although the value of a product is conditioned by its physical properties, only when a customer uses it is the value realised.

Applying S-D Logic to tap water implies that utilities don't just sell cubic metres of water but they facilitate experiences that consumers create by using the water. Water is not only essential to survival but also plays an important role in the social lives of people. The volume of water delivered to customers enables them to create experiences in their life. The physical properties of the water influence the experience of the customer, but the value of the water is more than the sum of its physical properties.

The framework has been developed into a canon of five axioms and six foundational premises (Vargo & Lusch, 2004, 2008, 2016). Although, from a marketing perspective, these statements can be considered axiomatic and not needing further proof, each of these is underpinned by research in sociology and economics. The study of material culture shows how people derive meaning from objects and how these objects play a role in their social reality (Appadurai, 1986). The most recent versions of the five axioms of Service-Dominant Logic are:

(1) Service is the fundamental basis of exchange.
(2) Value is co-created by multiple actors, always including the beneficiary.
(3) All economic and social actors are resource integrators.
(4) Value is always uniquely and phenomenologically determined by the beneficiary.
(5) Value co-creation is coordinated through actor-generated institutions and institutional arrangements.

1.1.2.1 First axiom

The first axiom declares services as the fundamental basis of all economic exchange. This axiom bridges the divide between products and services in classical marketing theory. S-D Logic does not deny the existence of tangible products and intangible services but it interprets both as distribution mechanisms of value.

A product such as water does not have any intrinsic value but delivers that value through its interaction with consumers. In traditional economic theory, value is created in the exchange of good or services, in S-D Logic this value is created through the use of the product. For water utilities this implies that, while volumes of consumed water and its chemical properties are still important, the subjective experience of customers is the act that creates value.

1.1.2.2 Second axiom

The second axiom states that the beneficiary is always a co-creator of value. Beneficiary is a generic term for customer, which generalises the application of S-D Logic to non-commercial contexts and internal marketing. A beneficiary can be the end-user of a service but can also refer to other stakeholders such as regulators, interest groups, or internal customers within the organisation.

Co-creation of value only occurs during consumption, mediated by the product on offer. Value is created in the interaction between the service provider and the customer, without which there is no value. From this axiom it follows that the service provider cannot produce intrinsic value, but can only offer value propositions that need to be accepted by beneficiaries for value to be created.

For water utilities, the axiom has implications for the way the value of water is understood and communicated. The value of water is more than its essential role in sustaining life. Water plays an important role in the social life of a water utility's customer. This understanding of the value of water is further discussed in Chapter three.

1.1.2.3 Third axiom

The third axiom concerns the status of the actors in the process of value creation. This axiom positions all economic and social actors as a resource integrator. This axiom moves beyond customer centricity and recognises the reciprocal roles of the service provider, the customer and third parties in the value creation process. Contemporary thinking in marketing introduces the term 'balanced centricity' for the situation where the interests of all actors in the value-creation network are taken into account.

Total customer centricity could be detrimental to the financial viability of the service provider or to the natural environment (Gummesson, 2008). Achieving a state of balanced centricity means that the differences in perception of the value proposition held by the various stakeholders are minimal. S-D Logic recognises that resource integration occurs in all nodes of the value supply network. Service providers integrate resources to create value propositions and consumers integrate resources to obtain the benefits they seek.

For water utilities as public service providers, this axiom implies that not only customers but also regulators, community groups and even the natural environment itself are beneficiaries of the value creation process. This third axiom highlights the importance of internal marketing which can be best described as the process of meeting the needs and wants of employees, and delivering positive experiences for them to ensure customers are best served. This mechanism is discussed in Chapter four.

1.1.2.4 Fourth axiom

The fourth axiom denies the existence of intrinsic value and places it firmly within the subjective sphere of the beneficiary. Phenomenology is the study of experiences from the first-person point of view. The beneficiary can be an internal customer, a regulator, a consumer or any other actor in the value-creation process. This axiom is the oldest truth in marketing: 'The customer is always right'.

Measuring the experience that customers have with a service is a complex concept. In the case of tap water, both the scientific view of the utility and the subjective view of the customer need to be taken into account (balanced centricity). Chapter five introduces SERVAQUA which is a statistically validated survey method to measure subjective service quality for water utilities.

1.1.2.5 Fifth axiom

The last axiom seeks to extend the perspective of marketing beyond the firm-customer model of exchange. This axiom invites researchers and practitioners to view organisations within their wider service ecosystem. S-D Logic goes beyond the traditional linear supply chain model from service provider to the customer. Services are provided through a value-creation network. This is a network of actors that includes all stakeholders, including suppliers, regulators, consultants, environmental groups and so on. The needs and wants of all actors in the network need to be satisfied in order to be truly customer-centric.

The value-creation network for water utilities includes many parties, with the paying customer and consumer of water as the final beneficiary. Chapter six discusses some of the issues related to managing relationships with customers and the community.

1.1.2.6 Applying service-dominant logic

S-D Logic provides a mental model for practitioners that helps them to filter their attention which guides the organisation's decisions and behaviour (Strandvik *et al.*, 2014). The canon of S-D Logic has been widely discussed in marketing literature, with many conceptual extensions and amendments being proposed. However, research on the applicability and utility of S-D Logic in managerial practice is rare (Blazquez-Resino *et al.*, 2015; FitzPatrick *et al.*, 2013; Lamberti & Paladino, 2013).

S-D Logic is not a theory that describes how services *should* be provided but it provides an account of how economic exchange actually occurs. Although Vargo and Lush (2016) refers to the five propositions as axioms, they follow from research into the sociology and psychology of economic exchange. S-D Logic is useful as a foundation for water utility marketing because it helps scholars to understand the industry, and helps professionals shape reality to maximise customer value.

The S-D Logic canon has been conceptually applied to the provision of public services. A service-dominant approach to public services situates citizens as central stakeholders of policy and public service delivery processes and it recognises that their engagement in these processes adds value. By taking a public service-dominant approach, co-creation becomes an inalienable component of public services delivery that places the experiences and knowledge of the consumer at the centre of service design and delivery (Osborne *et al.*, 2013). This logic also applies to water utilities. Tanji *et al.* (2012) applied these principles to irrigation services and they showed how the value of this service goes beyond the physical benefits of the water and how value is co-created between the consumer and the water provider. This book is the first comprehensive application of S-D Logic principles to the water industry.

Applying S-D Logic to the practice of water utility management does not consist of a series of simple steps to improve services. The model provides a way to understand the reality of water utility services in a way that is different to the traditional product-oriented approach. This understanding can assist water utilities to develop business practices that are centred around consumers.

1.2 WATER UTILITIES AS PUBLIC SERVICE PROVIDERS

The management of public services has been, over the past decades, influenced by a wave of managerialism through the introduction of New Public Management. This management philosophy encompasses a shift from the Weberian bureaucratic civil service to an approach more like the private sector. The concept of 'public administration' has been replaced by 'public management' to express the changed relationship between providers of public services and consumers of these services. Public service providers are more often adapting the behaviours of commercial service providers to maximise the value they provide to their customers. Managerialism in the public service has also introduced marketing models developed for competitive environments. The general premise for the introduction of marketing in the public sector is that using private sector marketing techniques contributes to better government.

Applying marketing in the public sector has not been without criticism. Public sector professionals often perceive it as being too overtly commercial to be of relevance to organisations concerned with the delivery of services that are a public good instead of a private good (Laing, 2003). Marketing is often critically analysed as the driving force of wasteful consumerism and the progenitor of false needs (Marcuse, 1964). Public service professionals generally adhere to this view and perceive marketing as being too overtly commercial to be of relevance to organisations concerned with the delivery of services that are considered a social good instead of a private good (Laing, 2003). The application of commercial marketing techniques tends to be viewed as unethical by public service professionals due to the contrast between the social good ethos and the commercial objectives of maximising consumption (Butler & Collins, 1995; Ryan, 1991).

Given that public services are generally not positioned to maximise consumption, commercial marketing techniques need to be modified to be applicable to public services. Also the monopolistic nature of the water utility sector requires standard models of market theory developed for private industry to be modified. A specific model, tailored to the characteristics of the water utility sector, needs to be developed to enhance the standard utilitarian view with an approach that embraces the symbolic value of water. Services provided by public organisations display distinct characteristics, not shared by commercially provided services, necessitating the modification of commercial marketing models (Butler & Collins, 1995; Hood, 1991; Laing, 2003; Lamb, 1987; Ryan, 1991; Whelan *et al.*, 2010). The following sections explain the differences between private and public services to better understand how marketing applies to water utilities.

1.2.1 Water as a public good

Water is often positioned as a public service because the resource is argued to be a public good (Robson, 2007). The debate around whether water is a public good is complex because of the double meaning of the words 'good', which can be either an ethical statement or a market offering.

Water is an ethical good as recognized by the human rights statement of the General Assembly of the United Nations. Principle four of the Dublin Statement on Water and Sustainable Development recognises water as a human right, but classifies it as an economic good (Gorre-Dale, 1992). The United Nations acknowledge that the economic value of water assists in its sustainable management. The status of human right does not preclude water from being a commercial good because food, clothing and housing are formal human rights and at the same time highly commercialised commodities (Bakker, 2007; Hanemann, 2005).

Notwithstanding the ethical status of water supply, water resources cannot be considered a public good in the economic sense of the word. Public goods simultaneously provide benefits to more than one individual as they are non-rivalrous in consumption (Figure 1.1). Public goods are also non-excludable, because their consumption by one person does not exclude another person from consuming it simultaneously (Zetland, 2011). Water systems are not public goods because consuming water is rivalrous. Water consumed by one person cannot be used for the same purpose by somebody else, without first increasing its value through purification. Although all consumed water eventually returns to the environment, it will be in a state of lower value through a loss of purity and pressure. The water cannot be returned as such to consumers within the same system, unless the water is recycled via technological means. Access to water resources is, in many instances, also excludable as the water is stored within a technological system, such as reservoirs and pipes.

A natural body of water should be viewed as a common-pool resource as it is a non-excludable but rival good. Common pool resources are those in which exclusion of beneficiaries through physical and institutional means is costly and exploitation by one user reduces the availability resources for others. Empirical research shows that no one single type of property regime for common-pool resources works most efficiently, fairly, and sustainable. Besides control by government or private entities, common pool resources can also be successfully managed collaboratively by communities. To prevent a 'tragedy of the commons', where self-interest driven consumption depletes a common pool resource, cooperation of local users is required (Hardin, 1968; Ostrom, 1990, 1999). From a S-D Logic perspective this idea underwrites the principle of co-creation of value between all stakeholders in a water system (Axiom 2). The cooperation of local users necessary to prevent depletion of scarce resources evokes the fifth axiom of S-D Logic.

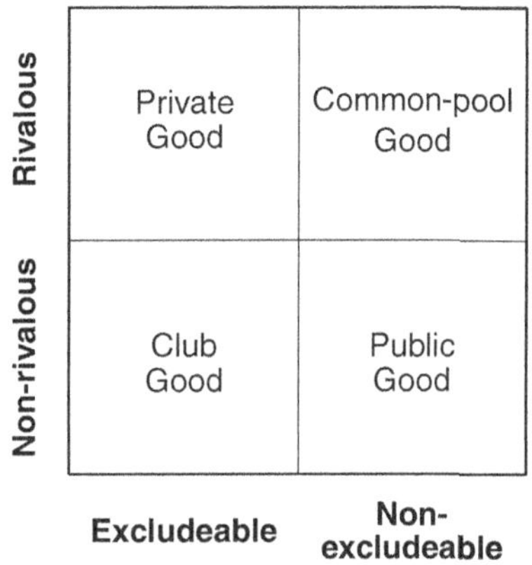

Figure 1.1 Classification of goods. *Source*: Cornes and Sandler (1986).

Successfully managing a common-pool resource requires personal constraint from individual users. Although water shortages are currently not a common characteristic of all water utilities, a large proportion of the globe is expected to experience water shortages in the near future, which will most likely require voluntary constraints in consumption by individual users (Macomber, 2013). Voluntary constraints in consumption are also a characteristic of public services in general (Butler & Collins, 1995). Water utilities manage private constraints by 'demarketing' consumption patterns. The importance of community involvement in the management of common pool resources, such as water, through voluntary consumption constraints places water utilities in the public sector sphere of management.

Water utilities temporarily convert the common pool resource of water into a private good by increasing its potential value through purification and pressurisation while storing it within a technological system that provides the service to consumers. This concept provides an economic justification for the criticism that water should be free because it falls from the sky (the public good). Water utilities do not charge customers for the common-pool good water but for the private good delivered by the assets required to purify and pressurise water. It is in this sense that water utilities are more like private service providers than a public service.

1.2.2 Natural monopolies

Customers generally have no choice of water supplier as the industry is a natural monopoly, which occurs when: 'due to the economies of scale of a particular industry, the maximum efficiency of production and distribution is realised through a single supplier' (DiLorenzo, 1996, p. 34). Natural monopolies exist because of the high level of investment required to duplicate water distribution networks, or the limited availability of water sources, creating an insurmountable barrier to entry for more than one firm to operate in the same geographic area (Hart *et al.*, 1997).

However, the current natural monopoly conditions are not a necessary condition. In the nineteenth century, water utilities operated in a competitive environment, with multiple firms servicing the same area. Utilities in the United States, France and England all started as private corporations (Goubert, 1989; Hunter, 1990). The functions of most of these private utilities were transferred to public ownership in the early twentieth century because they targeted wealthy areas and failed to invest in providing water systems equally among all citizens (Glennon, 2005). This change from private to public ownership was as controversial back then as the move to privatisation almost a century later (Bakker, 2007; Barnard & Henderson, 1917).

The past two decades have seen a gradual transition of ownership of utility companies, particularly communications, transport and energy, from public to private (Parker, 2012). Private corporations are now actors in about half the countries of the world in some key dimension of service delivery in the water utilities sector. Water services in, for example, France, England, several African countries and even communist Cuba are in private hands (Chisari *et al.*, 1999; Estache *et al.*, 2005). Notwithstanding this wave of privatisation, the majority of water suppliers around the world are still in public ownership. Irrespective of ownership, however, both publicly and privately owned water utilities operate as monopolies, with private companies licensed by government (Stevenson, 2006; van den Berg, 1997).

The human right statement of the United Nations and the Dublin Statement both acknowledge that access to improved water is essential for all human beings, including in those areas where it is economically a less attractive proposition. The recent wave of privatisation in water utilities has not universally improved access to water in poorer areas (Lee, 2011). Although some commercial water utilities use bottom-of-the-pyramid type strategies to extend networks in poor areas, there is no evidence that private ownership is better able to provide access to more people than public ownership of water utilities (Rangan *et al.*, 2011; Weidner *et al.*, 2010).

Natural monopoly conditions are strengthened by the fact that water utilities play an important role in prevention of disease. The recognition that water is important for public health started in the middle of the nineteenth century when it was recognised that cholera was a waterborne illness (Hrudey & Hrudey, 2004). The protection of public health is a core objective for all water utilities. However, the responsibilities for protecting human health are broader for water utilities than the product liability for other food manufacturers. There are no alternatives for tap water use because the community relies on the safety of the water for hygiene, hydration and cooking. This broad sweeping responsibility separates water utilities from other food manufacturers and places them within the sphere of the public service.

1.2.3 Ownership models

The criticism raised at water utilities for being inefficient and lacking customer focus have served as a justification to move control of this public service to the private industry (Auriol & Picard, 2009; Chisari *et al.*, 1999; Cunha & Cooper, 2002; Glennon, 2005; Hart *et al.*, 1997; Tan, 2012). Privately owned organisations maximise the return for stakeholders while the objective of publicly owned organisations are politically oriented. The argument developed in the literature is that a profit-maximising firm will improve quality to the point where the marginal benefit to the consumers of additional quality equals the marginal cost to the firm of increasing quality. For a politically oriented publicly owned firm, quality improvements are more related to the political will of the responsible governments than to customer perception (Fumagalli *et al.*, 2007).

Whether this logic holds in urban water services is disputable. Privatised water utilities also operate in a natural monopoly and thus demonstrate little intrinsic motivation to maximise customer value. Introducing effective management tools is not limited to privately owned utilities. New Public Management delivers improved public sector management by introducing private sector concepts and tools (Mukokoma & Van Dijk, 2013).

1.2.4 Public service characteristics

Public services can be placed along a continuum ranging from those dominated by social benefits to those characterised by a complex combination of social and private benefits (Laing, 2003). The distinction between social and private benefits relates to the respective roles of producers and consumers (Figure 1.2).

Social benefits dominant					**Private benefits dominant**
Customs & Excise	Criminal justice	Education	Health care	Public transport	Public housing
Professional judgement dominant					**Consumer judgement dominant**

Figure 1.2 Public services continuum. *Source*: Laing (2003). Reprinted by Permission of SAGE Publications, Ltd.

In services that predominantly provide social benefits, the offering is supply-driven, requiring professional judgement. In services that provide private benefits, the offering is driven by demand and shaped by consumer judgement. These distinctions can be used to guide customer service strategies in public sector organisations.

In services where the private benefits are dominant, the service is provided on a user-pay basis. In these instances the judgement of the consumer (the fourth axiom) dominates how the service is provided and what service is provided. Where social benefits dominate, there is no direct relationship between the individual customer and the service provider, and the views of individual customers are only taken into consideration through democratic political processes. In these instances, the professionals in the service organisation determine what service is provided and how it is delivered.

In services where public and private benefits are entangled, there is likely to be competition for dominance between professional and consumer considerations. The tension between professional and consumer judgement has been observed in health care where the individual demands of customers can be different from the medical needs of the patient (Bastian, 1998).

This tension also exists in tap water services. Urban water supply is a public service where private benefits of water use are dominant, but with a distinct aspect of social benefits through its role in reducing risks to public health. Tap water is sold to individual customers which renders the private benefits dominant. In simple terms, the customer defines what is valuable and what is not. The experience of the customer is defined by the way the water looks, smells and tastes, and the consumer's judgement dominates in that area. On the other side of the spectrum, the public health benefits are invisible to the customer. Tap water consumers are generally not able to assess the public health benefits of the water they consume, which makes professional judgement dominant in providing the service. The public service spectrum explains the tension that can exist between customers and water utility regarding the difference between the aesthetic and public health parameters of water.

Public services are dominated by political objectives rather than commercial objectives (Laing, 2003). Decisions in public service marketing are not necessarily based on commercial considerations (Hood, 1991) and are characterised by a prevalence of social justice and equity over profit maximisation: 'politics is fundamentally a moral undertaking and what is efficient comes second to what is right or good for the social community' (Walsh, 1994, p. 68).

Due to these differences in ideology between the public and private sector, the public sector is regularly characterised by demarketing rather than consumption maximisation, encouraging citizens to reduce the consumption of goods and services (Kotler & Levy, 1969), such as tobacco, alcohol and becoming involved with justice services. This is also the case for water utilities. In times of drought, shortages of water can force utilities towards demarketing and influence consumer behaviour in order to reduce water consumption (Lowe *et al.*, 2014; Walton & Hume, 2011).

Secondly, the relationship between service providers and customers in the public sphere is often limited by constraints. Public services, such as justice and education, require a different approach to customer relationships than commercial services. In services dominated by such constraints there is limited need for consideration of customer satisfaction as their defining purpose is the enforcement of compliance. In services less defined by these constraints, the use of marketing theory becomes more salient (Pasquier, 2012). Constraints in the relationship between water utilities and their customers depends on the jurisdiction in which the utility operates. An example of such a constraint are situations where water usage is restricted by legally enforceable water restrictions, constraining consumers to maximal values they can obtain from their water connection (Brennan *et al.*, 2007; Cooper *et al.*, 2011).

Thirdly, public services also distinguish themselves by limitations in the nature of the economic exchange between producers and consumers. A totally free exchange is considered to occur when both parties voluntarily exchange goods or services for money and when both parties believe they are better off due to the exchange. Most public services are provided by state owned monopolists, limiting the freedom of economic exchange (Butler & Collins, 1995; Lamb, 1987; Pasquier, 2012). Public services don't always offer defined services in exchange for a specific sum of money as taxation cannot be related to specific personal benefits. Public services in this category are not subject to market forces as there is no identifiable price that clears the market (Ryan, 1991).

Consumers of tap water have limited choices in alternatives to water supply from their local utility. The constraint in exchange is illustrated by the fact that in the water industry, water prices are in most cases regulated, with limited influence from the market on setting the appropriate rate (Brown, 2010a). The only exception is water wholesale where market driven forces set prices (Glennon, 2005; Phipps & Brace-Govan, 2011).

1.2.5 Summary

These considerations justify the idea that all water utilities, irrespective of governance structure or ownership model, are a public service. Although water is not a public good in the economic sense, the natural monopoly characteristics, the nature of the hydrologic cycle as a common pool resource, public health responsibilities and the essential nature of water implies that water utilities are more like public services than like commercial ventures. This conclusion does not imply that water utilities should be government-owned, but that the theories of public service management apply to this industry, irrespective of ownership.

This realisation implies that to develop a model for water utility marketing, considerations that hold for public service marketing also need to be applied to this industry. Marketing can be employed in water utility management as a tool to improve service provision.

For any marketing approach to be successful, it has to be developed specifically to the context in which it is applied. The characteristic differences between public services and commercial organisations warrants the development of discrete models for public sector marketing, matching the context of the sector under consideration (Butler & Collins, 1995; Laing, 2003; Ryan, 1991; Walsh, 1994). There is no one model for public sector marketing that can be applied across the diversity of services. Just as the private sector is a heterogeneous collection of industries, the public sector cannot be viewed as one and the same across sectors (Laing, 2003). A specific model for the marketing of water utilities is thus required in order to maximise value in this industry.

1.3 WATER UTILITY MARKETING

It is an immutable truth that water is an essential substance without which life is not possible. Water is also an irreplaceable resource for commercial processes, most significantly food production and power generation (Kurland & Zell, 2010; Schnoor, 2011; Wagner, 2013). Most discourse on the value of water revolves around these physical benefits. However, the intangible benefits of water are as important to human life as the tangible ones.

Although the essential nature of water would appear to be self-evident, it is also one of the least appreciated resources in contemporary society (Fagan, 2011). This limited appreciation is especially the case in the developed world where affordable, safe and reliable water services are available to every home as a matter of course. To increase awareness of the importance of access to clean water, the General Assembly of the United Nations declared that: 'the right to safe and clean drinking water and sanitation [is] a human right that is essential for the full enjoyment of life and all human rights' (General Assembly, 2010).

Water does not only enable physical life, but it also acts as a mediator for social interactions (Allon, 2006; Wagner, 2013). Creating a garden, bathing your child, doing the dishes with your partner or swimming with friends are mundane, but socially significant experiences. Water also has religious connotations, as it is used in baptisms, agricultural ceremonies and other types of rituals in cultures around the world (Lansing, 1987; Ten Bos, 2014; Wagner, 2013). In Australian indigenous cultures, water forms an integral part of the world of their ancestral beings conceptualised in myth, painting and dance. The practical aspects of water management are tightly integrated with spiritual life. Water is thus essential on three levels, to sustain life, enable economic growth and to mediate the social world (Barber & Jackson, 2011).

Air, water and food, the three essential elements of life, are in principle freely available in nature. Both food and water are supplied in sufficient quality and quantity through economic exchange. While air is freely available, to obtain food we rely on a complex network of producers, distributors and retailers. Water does not have to be produced like food; it follows the natural hydrologic cycle that starts in the oceans where it evaporates to form clouds and rain that subsequently nourishes rivers and groundwater. Water eventually evaporates or flows back into the ocean to start a new cycle. Water for human consumption is extracted from the hydrologic cycle, consumed and returned to the environment either as waste water, groundwater or through evaporation.

Besides being essential to life, water can also be a vector for disease. In proximity to human population and in its natural state, water is not suitable for consumption. Water is a universal solvent that often contains biological or chemical pollutants harmful to public health (Hrudey & Hrudey, 2004; Salzman, 2012). The World Health Organisation found that inadequate water supply is the direct or indirect cause of about ten percent of the global burden of disease. The incidence of waterborne disease can be significantly reduced by providing access to improved water services (Prüss-Üstün *et al.*, 2008). In most urban centres, water is therefore purified and pressurised and provided to communities by specialised service providers, the *water utilities*.

The International Organisation for Standardisation defines a water utility as the 'whole set of organization, processes, activities, means and resources necessary for abstracting, treating, distributing or supplying drinking water or for collecting, treating and disposing of waste water and for providing the associated services' (International Organizational for Standardization, 2007, ISO 24510). This definition encompasses all services within the economic sphere of the hydrologic cycle, including the collection of waste water from customers and recycling this either back into the environment or back to customers. In many service areas water supply and sanitation are provided by the same suppliers.

1.3.1 Water utility marketing literature

Supplying tap water to people has been discussed from many different domains of enquiry, including history, anthropology, philosophy, political science, engineering and the physical sciences (Strang, 2004; Ten Bos, 2014). Despite the criticism that water utilities lack customer focus, literature and practice about the application of marketing theory are embryonic both in industry and scholarly business literature (Kurland & Zell, 2011; Patsiaouras *et al.*, 2015).

1.3.1.1 Industry literature

The majority of peer-reviewed industry publications relate to the technical aspects of service provision, with limited reference to customers. There are several reasons for the limited presence of marketing scholarship in this industry. Water utilities are mostly controlled by engineers and scientists who view their vocation as a technological challenge, rather than a societal one (Allon, 2006;

Harrison & Stamp, 1991; Morgan & Smith, 2013). Water utility professionals do not consider marketing to be a natural partner because of the primacy of technology, their status as public service providers and the monopolistic nature of this sector.

Within the industry, marketing is commonly defined narrowly as synonymous with advertising and therefore superfluous to a monopoly service provider (Brown, 2010b; Harrison & Stamp, 1991). The attitude towards marketing in the engineering dominated water utility sector is illustrated through a textbook on business skills for utility engineers. This comprehensive tome on utility management excludes marketing as 'peripheral to most issues facing utility engineers' (Brown, 2010b, p. x). This exclusion is based on a misunderstanding of the scope of marketing as the text contains numerous references to customers, including issues such as willingness to pay and service expectations, which fall within the domain of marketing.

A range of industry-based research projects on service-centric tap water provision has been published as grey literature (Burns *et al.*, 2011; CUAC, 2013; Hall Partners & Open Mind, 2011; Means *et al.*, 2008; Ridolfi *et al.*, 2012; Vloerbergh *et al.*, 2007). These investigations are mostly descriptive and do not develop an overarching theory for tap water service provision. The industry literature contains some peer-reviewed studies on the application of marketing within water utilities but these consist of case studies or conceptual works, without the development of a theoretical framework (Addo-Yobo & Njiru, 2006; Coates *et al.*, 2004; Goetz, 2014a, 2017; Huberts *et al.*, 2000; Iacullo & Gradilone, 2000; Ruas Dos Santos, 2000).

1.3.1.2 Business literature

The monopolistic conditions of this industry have rendered it all but invisible to marketing researchers and the literature on water services within the domain of marketing is sparse (Kurland & Zell, 2010; Patsiaouras *et al.*, 2015). Economists have undertaken a significant amount of research into this industry, mainly related to pricing, efficiency and wholesale water markets (Estache & Rossi, 2002; Glennon, 2005; Grafton & Ward, 2008). The body of economic research includes an extensive corpus of publications on the impact of privatisation on the levels of service, which is also of interest from a marketing perspective (Auriol & Picard, 2009; Chisari *et al.*, 1999; Cunha & Cooper, 2002; Hart *et al.*, 1997; Rexha *et al.*, 2000; Tan, 2012).

Kurland and Zell (2010) conducted a systematic review of 49 top-ranked business journals and classified 135 relevant articles to develop a taxonomy of water-related business research. This taxonomy consists of six dimensions: *Water Quality* asks whether the water is clean or polluted. This dimension encompasses research on whether water is fit for purpose and discusses the perceptions that consumers have of the quality of water. The *Water Quantity* dimension relates to its availability and asks how we can develop resources or limit consumption. This dimension includes studies related to the amount of water used by consumers and how they can be motivated to moderate their consumption. The *Water Consumption* dimension refers to the benefits consumers enjoy from water. *Resource Management* centres on the sustainable use of water resources. This dimension is similar to the water quantity dimension, but views water use from the perspective of the water utility, instead of that from the consumer. Lastly, the *Company Management* and *Industry Management* dimensions centre on managing a water utility or issues involving the whole of the industry.

The scope of the taxonomy is limited to water used by businesses as a resource in economic activity, with limited reference to domestic water use. The review by Kurland and Zell (2010) also excluded marketing and economics journals. A further systematic review by Prevos (2016) of scholarly business literature extended the original taxonomy by including 141 marketing journals. This review found that the specific nature or marketing required an additional category for studies that discuss the relationships between utilities and customers, including service failures and service quality (Prevos, 2016).

1.3.1.3 Customer relationships

The privatisation of water utilities in the United Kingdom raised the level of attention placed on customers. The marketing strategy of Wessex Water is discussed in two case studies, with one paper announcing plans to become more customer focused and a later paper reporting on the achievements (Barrett, 1992; Skellett, 1995). Customer focus is often seen as superfluous for monopoly organisations, but it is nevertheless an important business driver. Satisfied customers complain less, which reduces unnecessary work. Satisfied customers are also more inclined to pay their bills promptly, resulting in improved financial business performance (Barrett, 1992).

Minimising the amount of customer contact is also beneficial for customers. The requirement for customers to contact the utility should be avoided to achieve internal efficiency and maximise convenience for the customer (Skellett, 1995). Customers of Wessex Water contact their utility only once every four to five years about a billing problem and once every ten to fifteen years about a service problem. This low level of interaction causes customers to take the service for granted, but they nevertheless expect a high standard of service. Contact mostly occurs after service failures and adequate management of service failures is, therefore, a critical moment of truth for water utilities. Careful management of these rare interactions ensures that after customers raise a problem with their utility, their appreciation of the service provider usually improves, even after a major incident (Barrett, 1992).

Sydney Water has also been the subject of a case study on their programme to increase the level of customer orientation (Harrison & Stamp, 1991). The board of Sydney Water realised that they no longer dealt with 'ratepayers' but with 'customers', people with 'legitimate concerns about the level of service' (Harrison & Stamp, 1991, p. 128). Their change programme moved the organisation from 'providing what the board thinks that customers need' to an 'informed understanding of customer's needs' and from 'passive customer contact' to 'proactive, courteous, helpful, friendly customer relations' (Harrison & Stamp, 1991, p. 130). There was initial resistance within the organisation to this change in corporate culture as marketing was perceived as synonymous with advertising and superfluous to a monopoly service provider. Sydney Water modified the marketing models developed for commercial organisations to suit the public sector characteristics of their industry.

What these case studies have in common is that both the British and Australian experiences with introducing customer-centric service provision were motivated by changes in legislation and regulation, in the absence of competitive pressures. These case studies demonstrate that privatisation is not necessary to create incentives for cultural change. New Public Management (NPM) is a model that has been used in water utilities to improve market orientation while the public sector retains ownership. The NPM paradigm states that public services can be improved by implementing private sector principles and practices. A case study conducted in Uganda reviewed the NPM approach as a method to increase the performance of a water utility along a range of indicators. They found that reform based on NPM can lead to improvement in technical efficiency and customer orientation (Mukokoma & Van Dijk, 2013).

1.3.1.4 Service failures

Given the high level of reliability and the essential nature of this service, any disruption in services constitutes a service failure and can impact large sections of the population. The manner in which utilities respond to such failures provides insight into the relationship between employee behaviour and service quality.

Major incidents, such as the Walkerton *E. coli* contamination and the detection of cryptosporidium and giardia cysts in Sydney are high-profile examples of high-impact service failures (MacGillivray, 2014). Another high-profile incident occurred in Northern Ireland. After a severe winter and rapid thaw, many water pipes in the service region of Northern Ireland Water (NI Water) ruptured, causing extensive disruption of services in many areas. This large-scale service failure caused anger and frustration in the community. The public discourse on this incident worsened the situation as NI Water attempted to shift the blame onto their customers. The water utility focused on the legal aspects and failed to build relationships with customers to manage this emergency (McCoy, 2014).

Both case studies show that the attitude of water utility employees hold towards Service Quality can reduce a customer's perception of the service failure. During water quality incidents, reports by consumers were ignored, until confirmed by laboratory evidence. The utility justified their inactivity to resolve the problem by referring to rules and regulations, without directly engaging with their customers on resolving the problem (MacGillivray, 2014; McCoy, 2014).

1.3.1.5 Service quality

A last substantial topic in the marketing literature is the quality of service provided by water utilities (Babakus, 1993; Burton *et al.*, 2001; Das *et al.*, 1996; Wirtz, 1998).

The earliest reference in this research area is a multi-item perceived Service Quality scale specific to electricity, gas and water utilities, based on the SERVQUAL model. This scale consists of fifteen items that are mostly related to supplementary services, such as billing and service requests, without considering the perceived quality of the core service (Babakus, 1993).

Burton *et al.* (2001) qualitatively assessed service quality in water utilities, finding that in services dominated by tangible elements, such as tap water, corporate image influences transaction satisfaction in addition to product and service quality. These findings imply that customer satisfaction is influenced by evaluations based on both direct experience with the service and information received from external sources.

1.4 SYNOPSIS

This book does not provide a simple step process for water utilities to follow to improve their organisation. This book provides a theoretical framework that practitioners can use to shape their thinking about their activities. S-D Logic is a mental model for water utility marketing that can be used to shape the decisions of water utilities. The S-D Logic framework does not prescribe a course of action but provides a model to describe the process of creating value.

Some of the detailed analyses that were used to derive these ideas have been omitted from this book. The dissertation on which this book is based contains a comprehensive analysis of S-D Logic from a theoretical perspective and provides the methodological justifications and statistical analysis. Readers interested in these deliberations can read the dissertation on which this book is based (Prevos, 2016). The dissertation is available from the author or the La Trobe University online research repository.

1.4.1 Sanitation services

This book focuses on the provision of water services and excludes sewerage or sanitation services. Although many water utilities provide both water and sanitation services, they are very different from a marketing perspective and should be discussed separately. From a marketing perspective, sanitation services are an extension of tap water services. Following the concept of product stewardship, all actors in the life cycle of a product (manufacturers, retailers, users, and disposers) share responsibility for reducing environmental impacts (Lane & Watson, 2012; Puffer & Al-Musallam, 2007). Product stewardship also applies to water utilities. Sanitation services not only improve public health; they also minimise the environmental impact of sewage.

1.4.2 Scope of this book

This book describes the retail of drinking water to customers in urban areas, delivered through networks of underground pipes. Other types of services provided by water utilities, such as sanitation and recycled water are outside the scope of this research. The reason for this focus is that water and sanitation services have different characteristics which require specific considerations.

Issues associated with the sustainability of water resources are also not considered. This book adheres to an economics-based view where demand for water is subject to the willingness and ability of people to pay, and which achieves equilibrium through price mechanisms (Grafton & Ward, 2008; Hoekstra, 1998; Meij *et al.*, 2005; Zetland, 2011). The uninterrupted availability of water is inextricably linked to the level of service. The sustainability of urban water supply can be defined as 'the development and maintenance of a required flow of benefits to a particular group or place, *undiminished over time*' (Gleick, 1998, p. 573, emphasis added). Nevertheless, the ability of water utilities to supply the required volume can be diminished in times of prolonged drought.

Any inability of water utilities to provide water at the required flow, for example through the legal enforcement of water restrictions, is considered to be a service failure (Cooper *et al.*, 2011). Although the environmental importance of a sustainable supply of water cannot be overstated and is a necessary condition of water utility management, it is not a sufficient condition to maintain a well-managed water system. In this book, environmental sustainability is considered to be a dimension of marketing as uninterrupted supply is a factor of the quality of service.

1.4.3 The water utility marketing mix

The next chapter analyses the relevance of marketing to the water utility sector and develops an industry-specific marketing mix. A marketing mix is a description of the activities needed to deliver services to customers and other beneficiaries. This marketing mix is defined by analysing the industry from a marketing perspective. A large sample of journal articles published by the International Water Association (IWA) were analysed to assess the relevance of marketing theory to this industry. The selected IWA journal abstracts were assessed for relevance and the identified topics were related to each other using a semantic network. This analysis revealed four communities of enquiry, which were used to define a marketing mix for this industry.

The water utility Marketing Mix consists of the *Value Proposition*, *Internal Marketing*, *Service Quality* and *Customer Relationships*. The Value Proposition is an expression of how the service provider positions itself in the mind of the customer. Internal Marketing entails all activities within the organisation undertaken to deliver the Value Proposition. Service Quality is an expression of the value provided to customers. Customer Relationships are most salient when service failures occur, or when customers require information or additional services. The remainder of this book discusses the four dimensions of water utility marketing in more detail.

1.4.4 The value proposition of tap water

Chapter three develops the value proposition for water utilities by identifying the determinants of value for consumers. This chapter systematically reviews literature on the anthropology of water consumption. The literature review is corroborated by interviews with Australian industry regulators and organisations that act as advocates for water utility customers. The research shows that, in line with S-D Logic, the determinants of value for water utility customers are not related to the physical properties of the water, but in the intangible benefits that customers enjoy.

A value proposition is not a mere slogan that succinctly summarises a promise from the utility to the customer. A value proposition is a comprehensive description of the value that an organisation creates with its customers and other actors in the value creation network. This chapter closes with a template to assist water utilities to develop their own value propositions. This value proposition can be used to anchor decision processes to make them more focused on customers.

1.4.5 Internal marketing

Chapter four investigates the crossroads between management and marketing. Internal marketing interprets employees as customers internal to the organisation and uses marketing techniques to enhance the value provided to employees. The ultimate aim of internal marketing is encouraging employees to implement the organisation's marketing objectives.

Organisations that rely on technology often experience tensions between the engineers that deliver the products or services and the employees that deal directly with customers. This chapter delves deeper into the role of water utility engineers and their relationship with customer-facing teams.

A strong relationship between engineers and customer service teams is essential in the effective delivery of value propositions. The research shows that tensions exist between engineers and marketers. The last section of this chapter introduces techniques to increase interfunctional coordination and engender a unified understanding of the value proposition.

1.4.6 Measuring the customer experience

The value proposition defined in Chapter two is management's view of the value provided to customers. However, the customer is always the ultimate arbiter of value. The experiences that customers have with their water utility can be assessed by measuring service quality. Chapter five describes a survey instrument to measure service quality in water utilities, based on S-D Logic. The model was tested using qualitative and quantitative data collected from water utility customers in Australia, the United States and Malta.

This chapter conceptualises service quality in two dimensions: scope and perspective. The scope of S-D Quality includes a technical and functional factor. Technical quality describes the core service and functional quality describes how the core service is provided. This chapter also introduces an index that summarises water system performance by combining technical data and customer complaints.

1.4.7 Customer relationships

Value is not created by the service provider with the customer passively consuming this service but co-created between both parties. To enable co-creation, water utilities need to understand their customers and manage the relationship they have with them. Chapter six discusses various aspects of customer relationship management as it relates to water utilities.

The first section defines the customers of water utilities by describing the value creation network. This chapter also analyses the various labels that are used to denote the customers of water utilities. This approach demonstrates that the customer definition for water utilities extends beyond the paying customer and includes many other stakeholders. The remainder of the chapter discusses traditional customer relationships.

This chapter also discusses customer segmentation and consumer involvement. Segmentation is a foundational strategy of marketing used to maximise value that is not often used by water utilities. Practical wisdom suggests consumers have a low level of involvement with water services. This chapter shows this not to be the case and discusses how this finding impacts the relationship between utilities and their customers. This chapter provides suggestions on how to manage the customer experiences by mapping the processes that create value. The final sections also discuss some insights on how to best deal with customer complaints and it will be argued that complaints should not be viewed punitively but rather be received as a gift to the business.

Chapter six closes with reflections on the invisibility principle. Logic suggests that a perfectly functioning water utilities does not directly interact with their customers beyond their plumbing. This idea seems to contradict general principles of marketing which hold that a strong relationship with customers is beneficial.

If we accept this idea, then how can water utilities develop a relationship with its customers?

1.4.8 Implementing water utility marketing

The final chapter summarises the findings of the research and provides suggestions on how to best integrate marketing theory with the practice of managing a water utility. The relationship between theory and practice in management disciplines is often strained by managers relating more to 'common sense' then to theoretical first principles. This chapter suggests a model to interpret theories in management. The role of theory is more than just a tool to do things. Theoretical reflection also changes the attitudes of managers, which in turn leads to improved performance through enhanced understanding.

This chapter also provides some reflections on the four paradoxes of water utility marketing. The value paradox suggests that, although water is essential for life, it commands a very low willingness to pay. The involvement paradox is similar in that common wisdom suggest that consumers have a very low involvement with their tap water, even though it is essential to maintain their lifestyle. The water quality paradox holds that the water quality perceptions of utility managers and of customers are often diametrically opposed. The last paradox is the invisibility paradox, which suggests that a perfectly operating water utility has no interaction with its customers. These apparent paradoxes are valuable to understand the specific issues faced by water utilities.

A book on a topic as broad as water utility marketing will never be fully complete. This book closes with suggestions for further research to enhance the industry's knowledge of marketing and how to use this knowledge to maximise the value they provide to their customers, consumers and the community as a whole.

Chapter 2

A marketing mix for water utilities

One of the most popular concepts in marketing is the so-called marketing mix, which is essentially a lists of the type of activities that a service provider needs to undertake to meet the needs and wants of consumers. The most well-known model is the 'Four Ps' of marketing (Product, Price, Place and Promotion). The Four P model implies that a marketer needs to develop a *product* for the right *price*, deliver this to the appropriate *place* and *promote* its existence to potential customers (McCarthy, 1964). Since its first inception, many alternative models for the Four Ps have been proposed. These alternative models either add more types of activities to the marketing mix or propose an entirely different point of view.

Many scholars and practitioners have criticised these models as being too simplistic as they place the firm instead of the customer at the centre of the process. The generic approach has also been criticised as not being able to account for the specific characteristics of the industry in which the organisation operates. In response to this criticism, several models have been proposed for other industries related to hospitality, internet commerce and wine retail (Constantinides, 2002; Festa *et al.*, 2016; Renaghan, 1981; Van Waterschoot & Van Den Bulte, 1992).

Even though many managers have been exposed to this idea through their training, the marketing mix has not found much application within the water industry. This chapter proposes a bespoke marketing mix for water utilities, based on their characteristics as service providers. Water utilities are public service providers that operate as a natural monopoly, which requires a specific approach to the marketing mix.

Peter Prevos
doi: 10.2166/9781780408675_027

The water utility marketing mix was developed by analysing a large sample of journal articles published by the International Water Association. This content analysis reviews the body of knowledge for urban water supply through the lens of marketing theory. Although these articles were not written from a marketing perspective, they could nevertheless be viewed within the context of marketing through their focus on customers. This analysis revealed the salient dimensions of water utility marketing as the *value proposition*, *internal marketing*, *customer relationships* and *service quality*. The water utility marketing mix forms the basis of the remainder of this book as each dimension is discussed in the following chapters.

2.1 THE MARKETING MIX

The concept of the Marketing Mix was first proposed more than half a century ago and stems from the idea that a manager is a 'mixer of ingredients' (Culliton, 1948, p. 5–6), who blends strategies and tactics to achieve organisational objectives. This metaphor led to the development of the Marketing Mix concept as an aid to thinking about marketing (Borden, 1964). In the original conception by Borden (1964), the marketing mix consisted of twelve items. This model was eventually simplified with the deceivingly simple mnemonic of *Product*, *Price*, *Place* and *Promotion* becoming known as the *Four Ps* of marketing (McCarthy, 1964). This version of the Marketing Mix has become synonymous with the discipline, propagated through the textbooks studied by future practitioners.

The Four P framework has not been immune from criticism. Since its original inception, numerous competing models have been proposed. Some scholars have added more Ps to the mix while others proposed entirely different models. One of the most cited extensions of the Four P framework is the extended marketing mix. This model was developed to allow for specific characteristics of services by adding three more aspects: *Participants*, *Process* and *Physical Evidence* (Booms & Bitner, 1981).

The *Participants* (or people) aspect recognises that consumption and production of services is inseparable and often requires interaction between people. The process of getting a haircut has very different marketing considerations then purchasing your razor due to the role that people play in the production process. In services, the customer and the service provider both have to be present, as the service is consumed while it is produced. The inseparability of production and consumption in services also emphasises the importance of the *Process* of service delivery. In the haircut example, the process of booking an appointment, the waiting facilities, the haircut itself, payment and so on, need to be carefully planned. The last additional component of the extended Marketing Mix states that *Physical Evidence* is required to mitigate the intangibility of services. Physical evidence plays a critical role in verifying that the service has actually been performed. Examples of physical evidence of a service are the programs sold at a theatre performance and souvenirs for tourists.

Applying this model is not always straightforward due to the generic nature of its seven dimensions. The extended model has been applied to water utilities in a marketing approach for water services in low and middle-income countries. This example demonstrates clearly that the authors used some creative license to fit each of the seven dimensions to the water industry (Coates *et al.*, 2004). Following the original interpretation of the model, the extended marketing mix could be applied to tap water as follows:

- *Product*: The physical quality of the water as measured by water utility.
- *Price*: The price of the water service paid by the customer.
- *Place*: The distribution system and the place where the water is consumed.
- *Promotion*: Any communication with customers.
- *Participants*: Employees that have direct or indirect contact with customers.
- *Process*: The process of connecting a new service, paying bills, moving to a new house and so on.
- *Physical Evidence*: The aesthetic properties of the water as experienced by the consumer.

This summarised list illustrates that some of the dimensions in the extended Marketing Mix don't fit comfortably within the water utility context. In most jurisdictions, the price of tap water is not the outcome of marketing strategies or the forces of supply and demand, but the result of a regulatory or political process. The vast majority of the moments of truth experienced by customers occurs in their homes. The water utility thus has little influence on the place of service delivery. Promotion is a rare activity for water utilities as there is no need to proactively acquire new customers. Most marketing communication relates to public relations, consumer education and managing emergencies. Furthermore, employees of water utilities rarely directly interact with customers.

This example highlights the need for a marketing mix specifically designed for water utilities that is developed in this chapter in two steps. The next section characterises water utilities as monopolistic service providers and discusses the impact of these characteristics on marketing water utilities. The following section presents a content analysis of water utility industry literature which will form the basis of the water utility marketing mix.

2.2 WATER UTILITIES AS SERVICE PROVIDERS

The supply chain for water services is a closed system that forms part of the hydrologic cycle. Utilities extract water from the natural environment and transfer it into a technological system where value is added and delivered to customers. Once this value has been consumed, customers return the water directly or indirectly to the natural environment. Water utilities add value to the hydrologic cycle through purification and pressurisation and deliver water directly to the customer's premises.

Purification increases the potential value of water by increasing the number of applications it can be used for. The number of uses is directly related to the purity of the water, with drinking water as one of the highest values of use. Pressurisation increases the potential value of water because it provides convenience to consumers by saving time in water use. The value chain for water utilities and its relationship to the hydrologic cycle is visualised in Figure 2.1. Each arrow signifies a process that adds potential value to the hydrologic cycle; the boxes symbolise the various types of core services provided by water utilities.

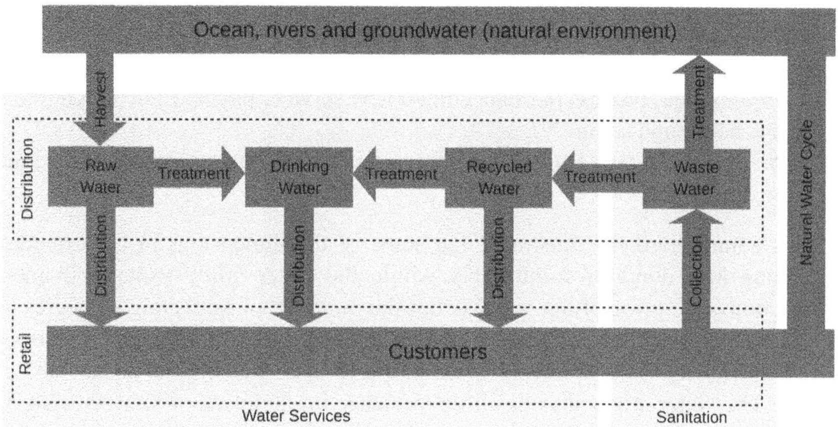

Figure 2.1 Water utility supply chain model.

Each type of water is suitable for different applications: raw and recycled water for irrigation, industrial and outdoor domestic use; drinking water for indoor domestic and commercial use. With respect to drinking water, the essence of the potential value added through this process is that water is safe to drink and pressurised. Without access to reticulated drinking water, consumers would need to invest knowledge, time and money to pressurise and purify water to ensure it does not pose a risk to public health (Vannini & Taggart, 2015).

Consumption of water is defined by a reduction in quality of the parameters salient to customers, namely the purity and pressure of the water (Lazarova, 2013; Zetland, 2011). Consumption does not destroy water but reduces its suitability for its original purpose. Afterwards, it is dispersed back into the natural environment at a reduced level of quality. In most developed countries, water is returned to the environment through sanitation services, which in many cases are also provided by water utilities. Water utilities can add value to the hydrologic cycle by ensuring that wastewater is purified to a standard that minimises harm to the environment.

Figure 2.1 also illustrates how waste water can be converted to drinking water through a series of technological processes, rather than relying on the hydrologic cycle. Technology ensures communities can reuse waste water, otherwise dispersed into the environment at reduced economic value. The reuse of sewage as potable water through technology is a controversial subject with profound marketing implications due to problems with community acceptance (Dolničar & Hurlimann, 2010; Dolničar & Saunders, 2006; Hurlimann, 2007; Kemp *et al.*, 2012; Menegaki *et al.*, 2009).

2.2.1 Tap water as a service

Unlike other raw materials, all water is ultimately returned to the environment and reverted to its original state through technological or natural processes. Once returned to the hydrological cycle, water can be included the next value chain. The urban water industry is, unlike other industries, characterised by this continuous loop between the environment and the economy (United Nations, 2012). In line with the first axiom of S-D Logic, the urban water supply should be analysed as a service, even though the market offering of utilities is the physical delivery of water. Consumption of water is transitory as the same substance is continuously used and returned to the natural environment. The water never changes ownership in a physical sense, the consumer purchases the right to consume water, not the physical substance itself. Customers purchase the right to use the water and pay for the value added by the service provider in the form of purification and pressurisation.

Water as a physical substance is never consumed but merely acts as a medium to transfer potential value in the form of purity and pressure to customers. Furthermore, consumers don't use water for the sake of using water but to create experiences. The value of a water service is co-created by the customer and the utility. The customer creates experiences with water, they shower, cook a meal, wash the dishes and so on. The utility enables these experiences by providing water at the appropriate level of purity and pressure while the customer seeks convenience, comfort and cleanliness (Strang, 2004). Although the water itself is a physical substance that is eternal, its value is perishable. Consumption of water extinguishes the potential value of water as its purity and pressure are reduced when it leaves the technological system of the water utility.

The market offering of water utilities hides a duality, as it can be both viewed as a tangible product and as an intangible service. Water utilities sell the water as a physical product, measured in widgets of water. The value experienced by customers is, however, expressed in moments of truth that cannot be measured as a physical parameter. The traditional marketing mix is not able to account for this duality. The marketing mix for water utilities thus needs to be able to account for both the intangible value of tap water and its physical properties.

2.2.2 Industry structure

Porter's Five Forces model (1979) allows for insight into the reticulated water industry. This model assesses the strength of competitive forces to help shape strategy. The Five Forces model describes competitive industries and might therefore not be suitable for water utilities. Vining (2011) proposed an alternative model specific to public services. In this model, the bargaining power of buyers is expanded with the power of sponsors who fund the service. The threat of new entrants is replaced with the influence of politics over the way the organisation is managed (Vining, 2011).

Although water utilities are public service providers (Section 1.2), Vinning's modifications are not required. Customers are the sponsors of their water utilities as the service is in most instances provided on a user-pays basis. The barriers to entry in the original Porter model also apply to water utilities because most water consumption is discretionary (Reed & Reed, 2011). The original Five Forces model defined by Porter is thus applicable to the tap water industry.

Research company MarketLine publishes reports on the tap water industry for 27 countries and regions, plus a global overview which includes Porter' Five Forces model. These reports target suppliers of goods and services to water utilities and for commercial organisations that seek to benefit from ongoing privatisation in this industry (Marketline, 2015).

2.2.2.1 Bargaining power of buyers

Water utilities are natural monopolies, eliminating customer choice over which service provider they prefer. This weakens the bargaining power that customers have with respect to the price of tap water. The weak power of water consumers is one of the driving forces behind the economic regulation of water utilities in many jurisdictions. Economic regulation prevents price gouging by water utilities and motivates service providers to be more focused on the needs of their customers (Franceys, 2006; Franceys & Gerlach, 2011).

The reliance on economic regulation to focus utilities on customers, plus the public ownership of most water utilities gives the industry an inherent political dimension. Regulatory agencies are established by the government and operate autonomously from the executive branch. However, the political dimension can dominate the industry as some regulators struggle to remain totally independent of government (Marketline, 2015).

Buyer power mainly exists through the political power of water consumers. Political power is expressed through democratic institutions and acts as a proxy for consumer power because the entire population within a service area is a consumer of tap water services. Intertwining customer satisfaction and voter endorsement is characteristic of public service provision, which is dominated by political, rather than commercial objectives (Laing, 2003).

2.2.2.2 Bargaining power of suppliers

Suppliers to the water utility industry are mostly providers of technological and scientific goods and services, as well as in some instances bulk water providers. The power of suppliers in the urban water industry is moderate because of their specialised nature of these supply industries (Marketline, 2015).

Although water utilities are mostly fully vertically integrated, many rely on the purchase of bulk water (Zetland, 2008). The power of these suppliers is high as only one source of water is available in most regions. Privatised source diversification through seawater desalination decreases supplier power for water utilities. The reliance on raw water suppliers is a dominating factor for the Public Utilities Board in Singapore. As a small country with limited area to capture rainwater, Singapore relies on Malaysia for the bulk of its water supplies, which gives much power to the supplier (Marketline, 2015).

2.2.2.3 Barriers to entry

While the undifferentiated services in other industries would motivate new entrants to invest, the natural monopoly conditions of the tap water industry are a disincentive to compete with incumbents (Marketline, 2015). The concept of a monopoly condition being natural is, however, not a necessary truth. DiLorenzo (1996) argues that natural monopolies are *ex-post* justifications for protecting government-owned water and energy franchises. The likelihood for competitors is dependent upon political will. In the Netherlands, Uruguay and Singapore, public ownership of water utilities is enshrined in law, preventing any competition for tap water services (Hall *et al.*, 2004; Marketline, 2015).

Energy services are a case in point to demonstrate that reticulated services are not necessarily natural monopolies. Many cities are serviced by both gas and electricity, both of which provide the same value to the consumer. The only reason people have more than one energy supply is because competition allows them to use the most suitable energy source. This example shows that the monopoly situation of water utilities is *de facto*, but not necessarily natural (DiLorenzo, 1996).

Water utilities might currently be considered natural monopolies, but these conditions are not certain to continue into the future. Visionary marketing scholar Theodore Levitt wrote a seminal paper about the concept of marketing myopia. This situation occurs when service providers are mostly inwardly focused and fail to recognise the higher-order value of their market offering. Levitt (1960, p. 47) proclaimed more than half a century ago:

> Who says that the [energy] utilities have no competition? They may be natural monopolies now, but tomorrow they may be natural deaths. To avoid this prospect, they too will have to develop fuel cells, solar energy, and other power sources.

These words were prophetic because energy providers that rely on reticulated networks are threatened by the proliferation of rooftop solar panels. Decentralised energy generation reduces the utilisation of high-voltage distribution networks, which negatively impacts the industry at a macro level (Robyns *et al.*, 2012). This grassroots development in the energy industry is driven by new technology and societal attitudes.

These attitudes can also be translated to water utilities. A small segment of the urban water market is seeking to live completely 'off the grid' and be self-sufficient for the supply of energy and water (Vannini & Taggart, 2015). The development of decentralised water systems is a possible threat to the monopoly currently enjoyed by water utilities. The further technological development of decentralised water systems, combined with a desire of certain customers to live independently could erode the economies of scale utilities currently enjoy.

2.2.2.4 Threat of substitutes

Although several alternatives to tap water, such as bottled drinking water and rainwater tanks, are available, these sources are not competitively priced compared to reticulated water (Francisco, 2014; Khastagir & Jayasuriya, 2011).

The above-mentioned decentralised water systems can be viewed as either a new entrant, if managed commercially, or as a substitute to tap water, if managed by the consumers themselves. In areas with unreliable water supplies, or those situated outside the servicing area of water utilities, private operators supply water to consumers through tankers (Ahmed, 2009).

Irrespective of these alternative services, the threat of substitutes is low because the total volume of these alternative sources is very small compared to the total volume supplied by the utilities (Marketline, 2015). In South Africa and India, the existence of traditional wells is interpreted as a moderate threat (Marketline, 2015). Furthermore, this source is not a suitable replacement for tap water as it carries a risk to public health.

Although there are no substitutes for water, there are substitutes to replace the benefits of outdoor water use, which accounts for about half of tap water consumption in Australia (Askew & McGuirk, 2004). Outdoor consumption is discretionary as consumers can choose to design their open spaces in a way that minimises or eliminates water use. During the Australian Millennium Drought, legally enforceable water restrictions spawned a cottage industry of alternative water use products, such as grey water systems that recycle water from showers and washing machines to maintain gardens. The drought also motivated many people to replace plants in their outdoor spaces with hard surfaces, reducing the need to use water (Allon, 2006).

Reducing water consumption through alternative benefits and sources has large economic consequences for water utilities. Water sales in the Netherlands have been negatively impacted by water conservation measures adopted by households and by alternatives to reticulated water, which forced water utilities to change their pricing structure (Meij *et al.*, 2005). The large ratio of fixed to variable cost caused by high asset depreciation implies that prices need to increase when consumption decreases. Zetland (2011) refers to increase of water prices due to decreased consumption as the 'utility death spiral'. This is a situation where revenues drop faster than cost and prices steeply rise, causing further reduced consumption.

2.2.2.5 Degree of rivalry

The degree of rivalry within the industry is low because incumbent utilities do not compete with each other for market share. In countries where licensed private companies supply services, the degree of rivalry is high due to the competition for licences (Marketline, 2015).

Although water utilities do not compete for customers, most regulators stimulate artificial competition. Regulators use market-mimicking systems through performance benchmarking to make utilities accountable to their customers. Under this approach, utilities compete with a shadow firm whose performance is determined by average or best practice in the industry . In this so-called 'yardstick competition', business performance of monopolistic service providers is compared with peer organisations (Braadbaart, 2007; Estache *et al.*, 2005; Lin & Berg, 2008; Tupper & Resende, 2004). Benchmarking can also be used collaboratively instead of competitively. In this approach, water utilities share best practice to improve performance of the whole industry and comparative performance standards are used as a proxy for commercial competition. Research in the Netherlands indicates that collaborative benchmarking enhances the economic performance of water utilities (Braadbaart, 2007).

The water utility industry is highly organised and information is freely shared at individual and organisational levels. The International Water Association is a global membership body that promotes information sharing between water professionals. At a national level, professional membership organisations, such as the American Water Works Association and the Australian Water Association, also facilitate sharing of information. Corporate membership industry associations, such as the Water Services Association of Australia and Vewin in the Netherlands provide advocacy for the industry as a whole. This high level of cooperation within the industry is enabled by the lack of competition.

2.2.2.6 Marketing tap water services

Analysing the forces within the industry reveals few differences between countries and shows that the structure of the water utility industry is essentially homogeneous around the world. Market forces in tap water supply are weak due to the monopoly in which utilities operate (Figure 2.2).

These weak market forces provide little intrinsic incentive for water utilities to maximise service levels. This pattern of industry forces is usually associated with high profit margins as business can exploit the low power of buyers, strong barriers to entry and weak threat of substitute products. Economic regulation acts as a substitute for competition by artificially increasing the bargaining power of buyers and simulating a higher level of rivalry between utilities. Limited economically viable substitute products are available for tap water. While the product of water cannot be easily substituted, the value it provides has many alternatives, particularly outdoor water use.

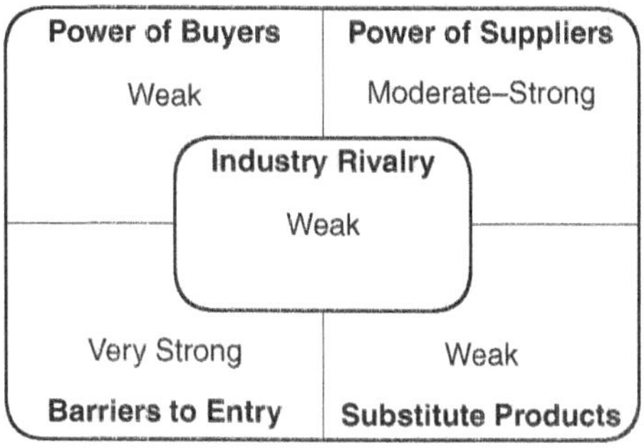

Figure 2.2 Five forces model for the water utility industry.

This industry review shows that, even though utilities enjoy the comfort of being a natural monopoly, there is a risk of marketing myopia, which could threaten the organisational sustainability of utilities. The barriers to entry in the tap water industry are very high but the emerging development of decentralised water systems could challenge the monopoly of utilities, placing pressure on their financial viability.

Pressure by regulators for more customer-centric service delivery, future threats to the industry through decentralised services, the drive for self-sufficiency of certain customer segments and the replacement of the benefits provided by water with other market offerings all support the idea of using marketing theory in this industry. Viewing water utilities from a marketing perspective assists managers to maximise the value provided to customers, lowering the desire for self-sufficiency and decentralisation.

2.3 CLASSIFICATION OF SERVICES

Services can be classified in many ways to describe the differences between the various industries. Classification helps to select the most effective marketing strategy as the nature of the service affects the way it should be marketed. To describe the differences between the various types of services, a range of classification systems have been proposed (Lovelock, 1983; Schmenner, 1986; Shostack, 1977; Stell & Donoho, 1996; van der Valk & Axelsson, 2015). This section reviews two of these classification methods to place water utilities within the wider context of the services industry.

2.3.1 Tangibility and intangibility

The importance of intangibility of services was first recognised by Shostack (1982) who proposed a sliding scale for services based on the dominance of either tangible or intangible aspects (Figure 2.3). All products and services consist of both tangible and intangible elements with different relative proportions of each. Products tend to be dominated by tangible characteristics whereas traditional services tend to be more intangible. Viewing market offerings this way resolves the duality between goods and services as there is no clear demarcation where services begin and products end. General advice in marketing is that intangibility should be avoided because it much harder to influence trust than to influence customers' perception of tangible elements (Shostack, 1982).

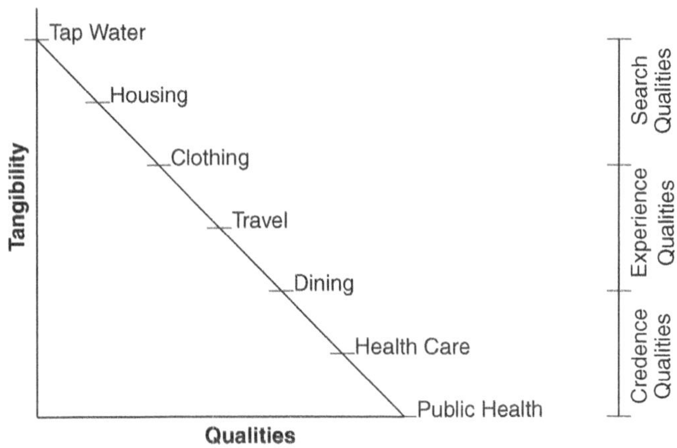

Figure 2.3 The tangibility continuum and customer evaluation.

From this perspective tap water is a service dominated by tangible elements. The tangibility of reticulated water renders it high in *experience qualities*, which means that customers can assess the level of service through their experience (Figure 2.3). Water can be seen, tasted, smelt, felt and heard by consumers, which provides tangible evidence of the service. The service nature of tap water lies in the fact that these experience qualities, the aesthetic appearance, can only be discerned during consumption.

Tap water also has *credence qualities*, attributes which customers find hard or impossible to evaluate, in part because they do not possess the knowledge or skills to do so (Darby & Karni, 1973; Rushton & Carson, 1985). Credence qualities in reticulated water are of particular importance regarding public health. This quality of water is intangible because it is not immediately apparent and difficult to evaluate for the average consumer as it can only be determined through laboratory testing.

This overview underscores the dual nature of tap water: both an intangible public health service and a tangible aesthetic product. This difference will become salient when discussing the measurement of customer experiences using service quality, in Chapter six.

Sanitation services are easier to characterise as they are defined by their complete lack of tangible elements. Customers don't want their sewer service to be perceptible. Sanitation only becomes tangible during emergencies such as sewerage spills. The lack of tangibility in sanitation complicates communicating its value to customers because the act of providing the service is invisible to them. Sanitation services, just like water quality, are high in credence qualities. These services rely on the customer trusting the professional judgement of the utility.

2.3.2 Service factories

The service process matrix developed by Schmenner (1986) amalgamates various other classification models and defines the managerial challenges for each type of service. The service process matrix has two dimensions: the degree of interaction and customisation, and the degree of labour intensity.

Following this model, water utility services are characterised by their low labour-to-capital ratio to provide a unit of service, which is the ratio of the total cost of staff over investment in equipment. The second dimension of water utility services is the low level of interaction between employees and customers. The core service is provided at the customer's premises through their plumbing and customers rarely need to interact with their utility. The nature of utility services also provides limited opportunities for customisation. All customers are connected to the same reticulation, which makes differentiation of products prohibitively expensive. An exception to this rule, some utilities in Australia have installed third pipe systems to deliver both drinking water and recycled water.

Services characterised by a low level of human interaction and customisation, and a low degree of labour intensity, such as tap water, are called *Service Factories* (Schmenner, 1986). Other examples of service factories are airlines, road transportation and accommodation. Examples of services with a high degree of interaction and options for customisation are hospitals, hair dressers, and repair services.

2.3.3 Classifying water utility services

The service process matrix and the tangibility scale distinguish tap water from other services. These two models can be combined to compare the services provided by water utilities with other service industries. In the classification model in Figure 2.4, the amount of customer contact and customisation, and the degree of labour intensity are shown on the horizontal axes; the level of tangibility is shown on the vertical axis.

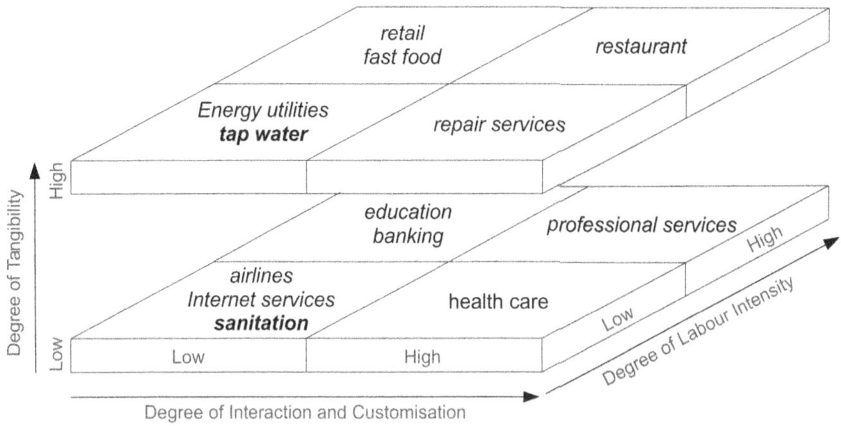

Figure 2.4 Service classification model.

Tap water services are similar to energy utilities, public housing and fast-food due to their low level of customer contact, high tangibility and low level of customisation. Sanitation services are more like telecommunication and local government services, because of their low level of tangibility.

The tangibility dimension is important to tap water because of its relationship to perceptions of quality through the difference between search, experience and credence qualities. The tension between the tangible and intangible aspects of tap water can lead to conflicts. In the absence of trust, consumers draw conclusions about the intangible aspects of water from its tangible aspects. In other words, customers with limited trust in their service provider are likely to judge the safety of the water on its aesthetic characteristics, that is its taste, odour and colour, rather than relying on the information from the utility. This tension can lead to the so-called water quality paradox, which is a situation when the health and aesthetic properties of tap water are in conflict with each other (Kot *et al.*, 2011).

The tension also relates to the continuum of public services (Figure 1.2). Regarding the intangible aspects of water quality, the professional judgements of the utility are dominant and utilities are in this respect a public service provider. With respect to the tangible elements, the judgement of the customer through their experience of the organoleptic properties of the water is dominant. From this perspective the utility is a private service provider.

The low degree of interaction and low labour intensity between water utility employees and its customers places special emphasis on those moments when an interaction does occur. The low level of interaction carries the risk of utilities losing focus on its customers as employees are removed from the experiences that customers have with their service. The low labour intensity implies a high reliance on technology to provide services. Organisations that rely on technology often experience tensions between the engineers and scientists that create the service and the customer-facing employees, which is discussed in Chapter four.

2.4 WATER UTILITY DISCOURSE ANALYSIS

The various marketing mix models are too generic to be successfully applied to water utilities. For a marketing mix to be effective, it should be grounded in the daily practice of management. Developing a marketing mix for water utilities can be undertaken by reviewing literature specific to this industry and analysing this literature in the context of its characteristics. The literature published by the water industry shows a unique snapshot of the discourse between professionals. Although this literature has in most cases not been written from a marketing perspective, it can nevertheless be viewed through the lens of marketing to gain insight.

The water utility sector publishes a large amount of literature on the specific problems faced by water managers. This knowledge is mainly disseminated in the form of reports commissioned by water utilities, research organisations, government agencies, or by commercial research organisations and universities. These publications are mostly classified as 'grey literature' as they are neither formally published nor independently peer reviewed. Studying this literature is problematic because bibliographic databases don't index these documents. Comprehensively reviewing grey literature is challenging due to the large number of repositories where this information resides (Mahood *et al.*, 2014).

The water industry also publishes peer-reviewed journals, with the American Water Works Association and the International Water Association as the largest and most influential outlets. These journals are accessible through databases and are open to systematic research.

2.4.1 Research method

The analysis was undertaken using all peer-reviewed journals in the English language published by the International Water Association (Table 2.1). These journals were searched on occurrences of the word 'marketing' in the title or abstract, plus a range of labels to denote users of water: 'community', 'consumer', 'customer', 'household', 'public', and 'user'.

This search strategy delivers only those articles that discuss matters related to customers and thus are likely to fall within the domain of marketing. Abstracts were chosen in lieu of full articles because only the topics of each article need to be identified, rather than a full analysis of the arguments contained within. Unsuitable entries were removed for a variety of reasons. Some entries contained

Table 2.1 English language peer-reviewed journals published by the IWA.

Journal	Scope	Since
Hydrology Research	All aspects of hydrology in its widest sense.	1970
Journal of Hydroinformatics	Application of information technology in the widest sense to problems of the aquatic environment.	1999
Journal of Water and Climate Change	All aspects of water science, technology, management and innovation in response to climate change.	2010
Journal of Water and Health	Research and practice across the full range of challenges to harnessing water for health.	2003
Journal of Water Reuse and Desalination	Theoretical and experimental research papers, new findings and issues of unplanned and planned reuse.	2011
Journal of Water Sanitation and Hygiene for Development	Information on the science, policy and practice of drinking-water supply, sanitation and hygiene.	2011
Journal of Water Supply: Research and Technology – Aqua	Scientific and technical, review, and practical/operational papers dealing with research and development in water supply technology and management, including economics, training and public relations.	1998
Water Policy	All policy aspects of water resources.	1998
Water Practice & Technology	Material that is of interest to practitioners rather than researchers.	2006
Water Quality Research Journal of Canada	Scholarly articles dealing with the aquatic environment.	2011
Water Research	All aspects of the science and technology of water quality and its management.	1967
Water Science and Technology	All aspects of the science and technology of water and waste water.	1982
Water Science and Technology: Water Supply	New developments in water supply.	2001

tables of contents, magazine articles or were written in languages other than English. Abstracts that discuss topics outside the scope of this research, such as bottled water, sanitation, agriculture or the wholesale of water, were also removed.

The broad definition of marketing promoted by the American Marketing Association (2013) covers the complete process of production, communication and supply of offerings. Using the broad definition of marketing enables the technical discourse in water utility journals to be assessed from a marketing theory perspective.

From a water supply perspective, this definition of marketing encompasses the value chain managed by water utilities, from extraction and supply of water to the provision of supplementary services. For example, articles about how customers perceive the chemical and biological properties of water can be viewed as discussing product quality. Articles which review means to promote behaviours to save water discuss. All topics discussed in water industry literature can in this way be mapped to concepts in marketing.

The abstracts were analysed using a method based on Grounded Theory and network analysis. The majority of the topics were sourced from two marketing dictionaries to provide a relevant perspective on the water industry literature (Doyle, 2011; Hart & Stapleton, 2012). Each abstract was assessed for best fit with any of the lemmas from the dictionaries. Any topics that could not be matched to a lemma in the marketing dictionaries, that is the design, construction, operation and maintenance of water supply systems, were coded generically as Asset Management.

The coded texts were further analysed to detect patterns within the discourse of the IWA literature. The relationships between topics define a semantic network,

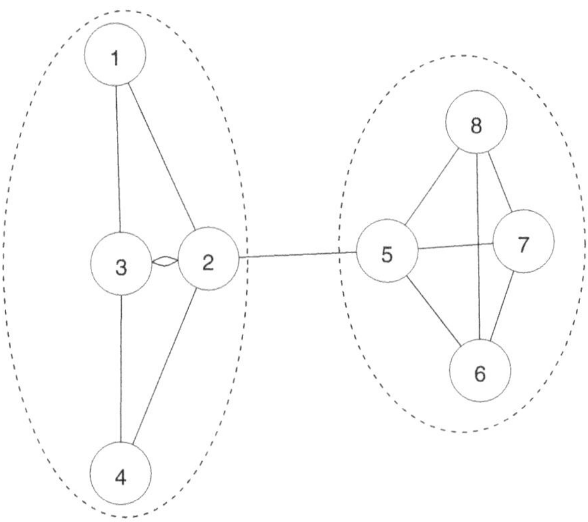

Figure 2.5 Example of a semantic network.

which can be visualised and analysed using standard network analysis techniques. An example of this approach is shown in Figure 2.5.

This example relates to four abstracts with eight topics. Abstract A describes topics 1 to 3, abstract B describes topics 2 to 4, abstract C describes topics 2 and 5, and finally, abstract D describes topics 5 to 8. These topics can be conceptualised as a network to visualise the structure of the discourse. Community detection algorithms are then used to detect patterns within the network of topics. In the example in Figure 2.5, the network shows two communities, indicated by the dashed ellipses.

2.4.2 Results

The thirteen journals were searched in May 2015 using the IWA and *ScienceDirect* databases, which contain over 40,000 articles. The analysed article abstracts are listed in Appendix A. The search resulted in 5,103 entries (Figure 2.6). Only the marketing, customer and consumer related search results were used to obtain a sample of the literature ($n = 590$). The other three keywords were excluded from the sample because they are too generic. The 590 search results were refined to a sample of 244 entries. Some abstracts contained more than one keyword, resulting in 225 unique journal article abstracts available for analysis. Only those articles for which the full text was reviewed are listed in the references (Table 2.2).

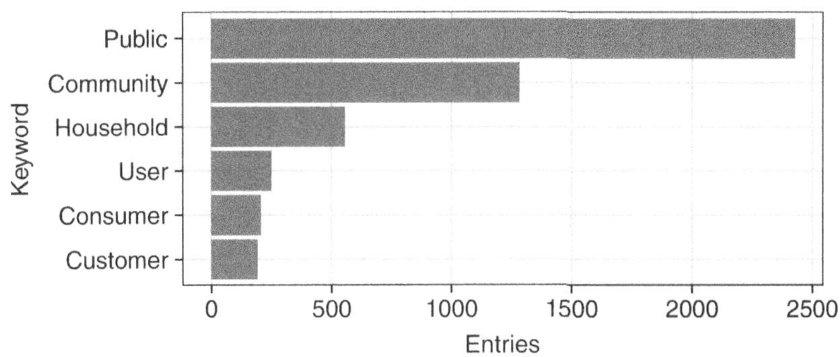

Figure 2.6 IWA journal search results.

The word cloud in Figure 2.7 explores the content of the collection of abstracts, excluding common stop-words and the search terms. The size of each word is proportional to its frequency within the corpus of abstracts. This diagram shows the prevalence of discussions regarding the quality of drinking water, specifically regarding its taste and odour (chlorine, geosmin). Words such as 'compounds'

Table 2.2 Results of the IWA journals search.

Journal	Abstracts	Consumer	Customer	Marketing
WST: Water Supply	67	27	44	8
Water Science and Technology	52	31	17	5
Water Research	33	25	8	2
J. of W. Sup., Research and Tech.—Aqua	23	9	13	1
Water Policy	17	11	9	1
Water Practice and Technology	17	7	10	0
J. of W. and Health	12	10	0	2
J. of Hydroinformatics	3	1	2	0
W. Quality Research J. of Canada	1	0	1	0
Hydrology Research	0	0	0	0
J. of W. and Climate Change	0	0	0	0
J. of W. Reuse and Desalination	0	0	0	0
J. of W. Sanitation and Hygiene for Dev.	0	0	0	0
TOTAL	225	121	104	19

Figure 2.7 Wordcloud of the IWA abstract contents, excluding stopwords and the search criteria (Fifty most common words).

and 'quality' indicate that the literature is dominated by research of systems and networks related to the treatment of water and how this impacts the performance of the utility.

A total of 27 unique topics were identified through an iterative process with a total of 591 topics assigned to the 225 abstracts (2.6 topics per abstract). The accumulated connections between all topics results in the discourse graph, visualised in Figure 2.8. A connection between two topics implies that they were both discussed in the same article. The larger the topic, the more it is connected to other topics. The graph is organised using the Fruchterman-Reingold method, which places nodes with the highest degree of connectivity in the centre and the lowest degree on the perimeter (Wasserman & Faust, 1994).

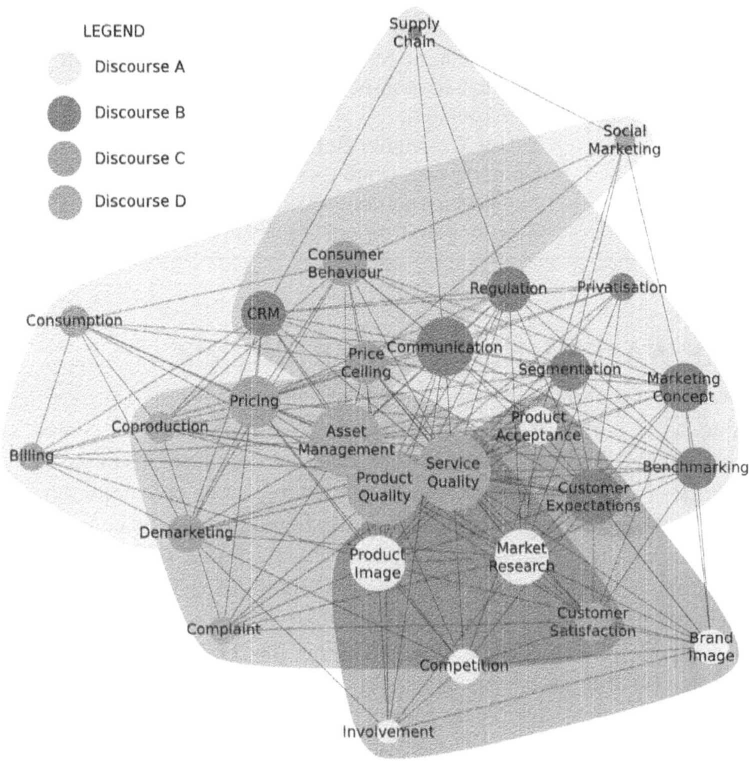

Figure 2.8 Communities of discourse IWA journals.

The Asset Management, Service Quality and Product Quality topics have the highest number of connections of all nodes in the network and form a prominent triad in the centre of the graph. These three topics dominate the discourse in water utility management. The least discussed topics are on the periphery of the network, such as social marketing, consumer involvement and supply chain management. Selective coding was undertaken by identifying communities using the Spinglass community detection algorithm (Newman & Girvan, 2004). The communities are outlined in Figure 2.8 in different colours and described in Table 2.3.

Table 2.3 Communities of discourse IWA journals.

Community of Discourse	Topics
A	Brand Image, Competition, Involvement, Market Research, Product Acceptance and Product Image.
B	Benchmarking, Communication, CRM, Customer Expectations, Marketing Concept, Privatisation, Regulation, segmentation and Supply Chain Management.
C	Billing, Consumer Behaviour, Consumption, Demarketing, Price Ceiling, Pricing an Social Marketing.
D	Asset Management, Co-creation, Complaints, Customer Satisfaction, Product Quality and Service Quality.

2.5 THE WATER UTILITY MARKETING MIX

The literature suggests that water utility managers have a tendency to view marketing as a peripheral activity (Brown, 2010b; Harrison & Stamp, 1991; Karbowiak, 2003). The content analysis of journal article abstracts published by the IWA provides further evidence of the predominately technological discourse in this industry. Based on a cursory reading of industry literature, it could be suggested that the marketing mix for water services is about *Pumps*, *Pipes*, *Plants* and *Processes*. However, this view is myopic of the customer's point of view as these aspects are invisible to them.

The network of topics that emerged from the IWA journals reveals four communities of discourse from which the water utility marketing mix emerges. Some of the topics have been moved between discourses to improve the interpretability of the results and to account for the characteristics of the industry.

The topics in the first discourse community mostly relate to the *Value Proposition*, which is the expression of value from the perspective of the customer. The second community of discourse relates to *internal marketing* activities that establish the utility as a customer-centric organisation. The third community consists of activities that build *relationships* between water utilities and customers. The topics in the last discourse community relate to *service quality*, both from the utility's and the customer's perspective (Table 2.4).

Table 2.4 Water utility marketing mix.

Marketing Mix	Topics
Value Proposition	Brand Image, Competition, Involvement, Market Research, Product Acceptance, Product Image.
Internal Marketing	Benchmarking, Communication, Marketing Concept, Privatisation, Regulation and Supply Chain Management.
Customer Relationships	Billing, Complaints, Consumer Behaviour, Consumption, CRM, Customer Expectations, Demarketing, Price Ceiling, Pricing, Segmentation and Social Marketing.
Service Quality	Asset Management, Co-creation, Customer Satisfaction, Product Quality and Service Quality.

These four dimensions are more or less in a causal relationship with each other. Management defines a value proposition that is communicated to employees through internal marketing. The employees of the organisation deliver services and manage relationships with customers. The experience that the customer has with these services is expressed in the quality of the service (Figure 2.9).

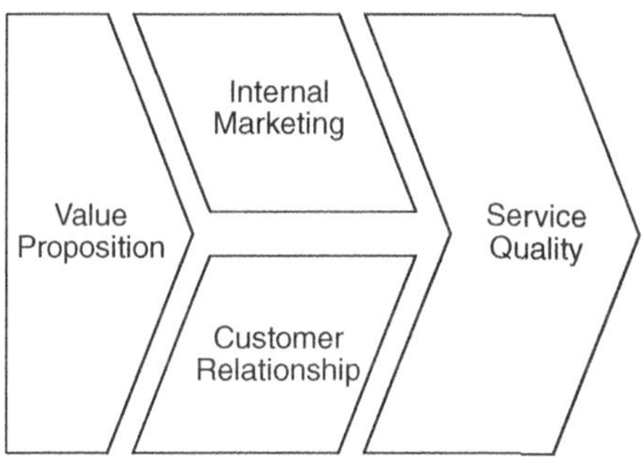

Figure 2.9 Water utility marketing mix.

2.5.1 Value proposition

The first dimension of the water utility marketing mix consists of topics that revolve around the value proposition. The topics in this community of discourse are mainly concerned with the perception that consumers have of the utility as a service provider. A value proposition is more than just a promissory statement, it is an expression of how the service provider positions itself in the mind of the customer and guides all further marketing activities (Anderson *et al.*, 2006; Frow & Payne, 2011; Payne & Frow, 2014).

Brand image, product image and product acceptance are closely related topics. Branding is not commonly adopted as a strategy in the water context, with the notable exception of bottled water. The brand image of a water utility expresses the physical quality of the product and influences the level of trust consumers have in the provider. The issues water utilities can face with the acceptance of alternative water sources are central to the product image and product acceptance discourse. For instance, the acceptance of drinking water sourced from waste water relies on the level of trust consumers have in the utility (Dolničar & Hurlimann, 2009; Dolničar *et al.*, 2014; Doria, 2006, 2010; Fife-Schaw *et al.*, 2008).

Another topic in the value proposition is the influence of absence of competition. Dupont and Jahan (2012) showed that consumers in Canada are willing to spend significant amounts to obtain drinking water that they perceive to be safer and cleaner than tap water. Defensive spending on tap water substitutes falls within the sphere of the value proposition because the shift in demand is motivated by a perception of the value provided by water utilities. The increased popularity of alternatives to tap water is not competition in the sense of a threat to market share, but an erosion of trust in the water utility's ability to provide sufficient product quality (Levallois *et al.*, 1999; Meij *et al.*, 2005; Puget *et al.*, 2010).

The IWA abstracts predominantly discuss the value of water within the context of its physical quality. One study in the IWA journals described the difference between water as a physical product and its symbolic value (Ruas Dos Santos, 2000), which is the view commonly held by anthropologists and corresponds with the first axiom of S-D Logic. The anthropological literature emphasises the intangible value of water and argues its importance over the material value of water. Only a minimal amount of water is required for survival, with most of it used for identity creation and cultural capital (Askew & McGuirk, 2004; Reed & Reed, 2011; Shove, 2003).

The value proposition of water does not reside in its physical properties but in the benefits obtained from its consumption, both tangible and intangible. The physical properties of water are an important aspect of the value proposition, but only to the extent that they transmit the use value to customers. This view on value propositions is discussed in detail in Chapter three.

2.5.2 Internal marketing

The topics in the second dimension of the water utility marketing mix relate to internal marketing because they mostly describe internal activities that indirectly impact customers. Internal marketing are activities undertaken by management to improve the level of service by ensuring that employees understand the value proposition.

In an absence of competition, benchmarking is a commonly used tool in the industry to measure and improve performance, which occurs both voluntarily and through regulation (Braadbaart, 2007). Benchmarking activities are focused on cost or physical quality but can also include supplementary services

(Iacullo & Gradilone, 2000; Merkel *et al.*, 2011). Supply chain management, from an engineering point of view, is one of the salient topics in the industry literature (Adams, 2008). The marketing perspective of this function is only recognised in the sense of demarketing (Ameyaw *et al.*, 2013; Barrett & Wallace, 2009; Zubizarreta, 2008). Balancing supply and demand is mostly a social science problem rather than an engineering problem (Erlenkotter *et al.*, 1989).

The application of marketing is slowly taking hold within the industry. Although the literature paints a negative picture concerning customer orientation in the industry (Brown, 2010b; Harrison & Stamp, 1991; Karbowiak, 2003) and industry literature focuses on technology over marketing, there are occasional mentions of a need for a market-oriented approach (Huberts *et al.*, 2000; Iacullo & Gradilone, 2000; Mugabi & Njiru, 2006; Mugisha, 2007). Chapter four discusses internal marketing with specific reference to the role of engineers within water utilities.

2.5.3 Customer relationships

The frequency of the six search words 'community', 'consumer', 'customer', 'household', 'public', and 'user' shows a clear preference for labelling the beneficiaries of water services (Figure 2.6). The more impersonal label 'public' is preferred by the industry over the more personal customer label. This indicates a preference for the public service nature of water utilities—to the community as a whole—over the individual customer approach.

Relationships between customers and water utilities are often based on service failures as customers usually only contact the service provider when expectations have not been met. This suggests that direct contact between the utility and the customer should be minimised (Skellett, 1995). Lowering the amount of contact implies lowering the number of service failures and thus increasing the level of Service Quality.

The IWA literature supports that tap water consumers have a low level of involvement, which seems to contradict the essential nature of this service (Howarth & Butler, 2004; Ruas Dos Santos, 2000; Wijkman & Gordon, 2002). Due to the low level of involvement, customers are not considered to be interested in a relationship with the basic service provider. Water utilities can, however, benefit from developing a positive relationship with their customers to improve the process of price determinations and demand management (Karbowiak, 2003).

Pricing and willingness to pay (the price ceiling) are problems for essential service providers. The willingness to pay for tap water is very low due to the water-diamond paradox. This paradox holds that some commodities without value in use, such as diamonds, have a high value in exchange and vice versa, water has a high use value, but demands a low price (Levy, 1982; Robertson & Taylor, 1957; Smith, 2010). Water is used for many purposes and in the case of reticulated water supply, the same service is used for drinking, hygiene, gardens and as a transport medium for waste.

Economic theory suggests that the price of a product is determined by its marginal utility, that is the least important use to a person (Levy, 1982). In the case of tap water, this is water's function as a transport medium for waste. There is no invisible hand guiding markets to regulate water utility prices. As public service providers pricing is a political issue, which places limits on the prices utilities can charge. Customer relationships are important in managing customers that struggle paying bills, which positively impacts the utility's cash flow. The sample of IWA literature discussed the topic of financial hardship mainly within the context of developing countries (Addo-Yobo *et al.*, 2006).

Discussions about complaints mostly refer to the physical quality of water through either discolouration, taste or odour issues (Boxall *et al.*, 2003; Dietrich *et al.*, 2014; Gallagher & Dietrich, 2014). Through the complaint management process, customers act as sentinels of the water supply system and inform the service provider of problems not detected by technological monitoring methods. Water services are a sensory experience but unlike other services good tap water is characterised by the absence of sensory stimuli, neither taste, colour or odour.

The last aspect of customer relationships with water utilities is demand management. In countries that suffer from large variability in water availability, the relationship between the utility and its customers is occasionally dominated by efforts to moderate water consumption through social marketing. Social marketing is a subset of demarketing, which are deliberate strategies to reduce demand (Cullwick, 1975; Kotler & Levy, 1971).

Water utilities regularly conduct market research related to taste and odour perceptions of water and overall customer satisfaction (Bruvold, 1970; Camps, 2000; Dietrich *et al.*, 2014; Howarth & Butler, 2004). In countries affected by drought, a major topic of interest are the community's attitudes to alternative water sources (Dolničar & Hurlimann, 2009).

The topic of segmentation is rarely discussed in the industry literature. Socio-economic factors and the related risk of financial hardship was the most common segmentation variable. This finding corroborates the information obtained from customer advocates whom all stressed the importance of specialised services for customers who struggle paying their bills (Freeman *et al.*, 2009; Puget *et al.*, 2010).

2.5.4 Service quality

The last topic is related to service quality, which is defined by customer perceptions about their experience with service providers. Given that the vast majority of the moments of truth experienced by customers are related to the core service and that these services are delivered through technological means, the discourse is dominated by deliberations on asset management. These articles discuss 'how best to manage … existing asset stock to provide satisfactory customer service with limited funds' (Burn *et al.*, 2010, p. 55). Or as Johnson and Mortada (2005, p. 97) expressed it: 'asset management is a critical link in balancing overall product or service delivery cost of ownership to achieve and sustain customer satisfaction'.

The debate on the quality of tap water services revolves around methods to measure and control the physical quality of tap water. Articles about service quality discuss the perceptions that customers hold of the colour, taste and odour of tap water and to a lesser extent the quality of supplementary services such as billing and providing information. Perceptions of water quality result from a complex interaction between factors such as risk perceptions, attitudes towards chemicals, familiarity with specific water properties, trust in suppliers, past problems attributed to water quality and information provided by the mass media and interpersonal sources (Burlingame & Mackey, 2007; Deichmann & Lall, 2007; Doria, 2010; Krishna & Morrin, 2008).

Perceptions of quality are complicated by the water quality paradox (Kot *et al.*, 2011), which is caused by the tension created by the private benefits of aesthetically pleasing water and the social benefits of healthy water (Parr, 2005). Although providing safe water is a necessary condition, it is not a sufficient condition to secure customer satisfaction. The taste and odour of drinking water are psychologically closely associated with public health, even though there is no physical relationship between them. Water free of objectionable taste and odour renders the value proposition of tap water tangible to consumers.

Service quality, as a way to measure customers experience, is analysed in detail in Chapter six. The chapter includes a psychometrically validated scale to measure service quality.

Chapter 3

The value proposition of tap water

Contemporary marketing publications often promote the virtues of crafting the value proposition for an organisation. This concept has attracted a lot of attention in the strategy and marketing literature. Kaplan and Norton (2004) called it the 'most important organising principle' for any business, and others have referred to the value proposition as the essence of strategy (Payne *et al.*, 2017). In its most simplistic form, a value proposition is interpreted literally as a promise of value delivered to the customer. The value proposition is often presented as a slogan, a statement that summarises a promise of value from the service provider to the customer.

Even though this definition is straightforward, many managers struggle with the concept. The main confusion with value propositions is that they are more than just promissory statements of expected service. The value proposition is a complex notion that can be viewed from different perspectives within the value-creation network. Value propositions inform all aspects of the process of creating value. This chapter proposes a model to assist water managers to understand and craft their utility's value propositions.

It is an immutable truth that water is essential to life. Water managers are astutely aware of this fact and descriptions of the value of water regularly use the superlative 'invaluable' to emphasise the essential nature of the services they provide (Roa *et al.*, 2014). Water is indeed, together with air and food, essential to sustain life but the value of water has more dimensions than mere survival. Water is used to live, for economic growth and to enjoy life. The consumption of water is not only

doi: 10.2166/9781780408675_053

essential to meet physiological needs, it is also required to meet the higher-order psychological and social needs.

Tap water is not merely a means to maintain personal hygiene. The act of bathing also serves to create our public self and provides a private moment of reflection (Rook, 1985; Shove, 2003). A mundane activity such as gardening, in which water plays a vital role, serves to meet the full spectrum of human needs, for example physical exercise, expressing creativity and creating a space to socialise (Solomon *et al.*, 2010). The further the consumption for water is removed from the need to survive, the more water is used to meet those needs. The self-actualisation provided by gardening or the water needed for the daily grooming rituals require much more water than what we need to survive (Reed & Reed, 2011).

The discourse of water consumption classifies some uses as necessary ‘needs’ and to other uses as discretionary ‘wants’. The marketer’s view of needs and wants is quite different from this common sense understanding. The need to survive, the need for social belonging, self-actualisation and other higher order needs, are of equal importance to physiological needs (Solomon *et al.*, 2010). The common interpretation of needs and wants is based on a value judgement. Something people need is considered more important than something they want. Defining needs and wants in this way is not effective in marketing because it forces the value-system of the service provider onto the consumer. The common sense view thus limits our ability to clearly define the value proposition of water. In Service-Dominant Logic (S-D Logic), value can only be defined by the consumer, from which it follows that the service provider should not promote their judgement of a perceived difference between needs and wants.

However, the specific characteristics of water utilities require special considerations in some cases. The tension between being a service provider and being a public service, discussed in Chapter one, occasionally forces water utilities to make value judgements about the way water is consumed. Water utilities around the globe are actively involved in shaping the way water is consumed and promoting water conservation efforts as part of their role as a public service provider (Lowe *et al.*, 2014; Walton & Hume, 2011).

This chapter is based on the idea that needs are innate to the human organism and range from survival to self-actualisation (Maslow, 1943). All these needs are equally valid reasons to use water. Wants are the socially constructed ways in which these needs are fulfilled. Thirst can be quenched by tap water, but also by other drinks, and most of our daily intake of water comes from the food we eat. A feeling of social belonging can be fulfilled through many ways, only some of which, such as a backyard pool, requires water.

Positioning the value of water as a multivariate parameter rather than emphasising its role in biological survival aligns with contemporary thinking in marketing and anthropology. A foundational truth in consumer behaviour is that people don’t buy things for what they do, but for what they mean. The value of a market offering, whether a fast-moving consumer good or tap water, always lies beyond its direct functional purpose (Arnould, 2007; Douglas & Isherwood, 1996). Needs and wants are not value judgements but conceptually different aspects of consumption. Understanding the relationship between needs and wants is essential in order to craft meaningful and useful value propositions.

In marketing, needs are defined as felt states of depravation that can be physical or psychological, all of which are necessary to lead a fulfilling life. Wants are the desires by which these needs are fulfilled.

Following these thoughts on needs and wants, a value proposition can be described as the confluence of the reason *why* people purchase services (the need) with *what* the service entails (the want). Commercial organisations use value propositions to differentiate themselves from their competition. Value propositions can assist service providers that operate in a non-competitive environment, such as water utilities, to position themselves in their community and towards their stakeholders. The purpose of the value proposition is to inform the marketing activities that water utilities undertake in order to align them with the needs and wants of their customers.

This chapter provides a practical framework to develop value propositions for water utilities. The first section defines value propositions and their role in the value-creation network. The following section describes the societal role of water utilities from an anthropological perspective. The science of anthropology is interested in the interactions of everyday life and is a fertile source of information for water utility marketers. This review of the anthropology literature is enhanced by interviews with Australian organisations that represent customers of water utilities. This chapter closes with a template to assist managers who want to develop a value proposition for their water utility.

3.1 VALUE PROPOSITIONS

Kowalkowski (2011, p. 277) defined value propositions as 'promises of reciprocal value between service providers and their customers'. This common definition is a literal interpretation of the term that only provides a limited view on the concept. A value proposition is more than a promissory statement of expected performance based on the consumers needs. In S-D Logic, the value proposition constitutes the traditional marketing mix (Product, Price, Place and Promotion). The S-D Logic value proposition consists of all aspects of the value delivered to customers and consumers, from the perspective of each relevant actor in the value-creation network. The value proposition is thus not just a statement in words, but it encompasses everything the service provider does that can be experienced by consumers.

Contemporary views on marketing are firmly grounded on the concept of value. Within S-D Logic, two of the four axioms anchor the marketing process to the creation of value. The co-creation axiom stipulates that value is always created in an interaction between the service provider and the customers. Value is not designed and provided in isolation by the service provider but in an interactive process between the various parties in the value-creation network. Customers don't purchase goods or services, but offerings that create value through their experience with them (Gummesson, 1995; Macdonald *et al.*, 2011).

Goods and services are not imbued with value or utility during the production and distribution process. Value also isn't created in the exchange of the goods and services, but only comes into existence when the product or service is experienced. Physical goods are, in the words of Lusch and Vargo (2014, p. 13): 'appliances that act as intermediaries in service delivery'.

Although water is delivered at 'arm's length' from the customer through their plumbing, the water itself has no value until it is used by the consumer. Value is not created in the water treatment plant or in the network, but in the customer's home. The value of tap water is co-created between the utility and the consumer.

The fourth axiom of S-D Logic refers to the old saying that the customer is always right, emphasising that they are the final arbiter to decide what is valuable (Lusch & Vargo, 2014; Vargo & Lusch, 2016). Each actor in the value-creation network has his or her own perspective of the value proposition. From a marketing perspective, the value of a service can only be assessed by the consumer. However, in tap water there is a tension between the aesthetic quality of the water as perceived by the consumer and the safety of the water as perceived by the scientist. Consumers might be dissatisfied with the chlorine or fluoride added to promote public health. Even though the benefits of these two chemical additions are beyond doubt from a scientific viewpoint, the customer's perspective can conflict with the perspective of the scientist, referred to as the water quality paradox. The customer's interpretation of the value proposition is dominant but the perspective of the water utility is also of importance. This tension is a recurring theme in water utility marketing.

The value proposition exists throughout the value-creation process and is interpreted differently by each of the actors in the value-creation network. Management defines the value proposition, employees deliver this and the customer eventually experience it. Optimal performance is achieved when customers experience the value proposition the same way as it was designed and delivered by the service provider. In optimal performance, there are no gaps between the management, the employee and the customer value proposition (Figure 3.1). This alignment of perspectives defines the very core of being a customer-centric service provider. In some instances other perspectives of the value proposition, such as that of regulators or community groups, also need to be accounted for.

The management value proposition, which is the subject of this chapter, is the formulation of value as defined by management. This value proposition is translated into an employee value proposition which functions to ensure employees are able deliver the value envisioned by management, as discussed in Chapter four. The ultimate arbiter of value is always the customer who co-creates value with the organisation. The customer value proposition is defined by the way the customer experiences the service, which is discussed in Chapter six.

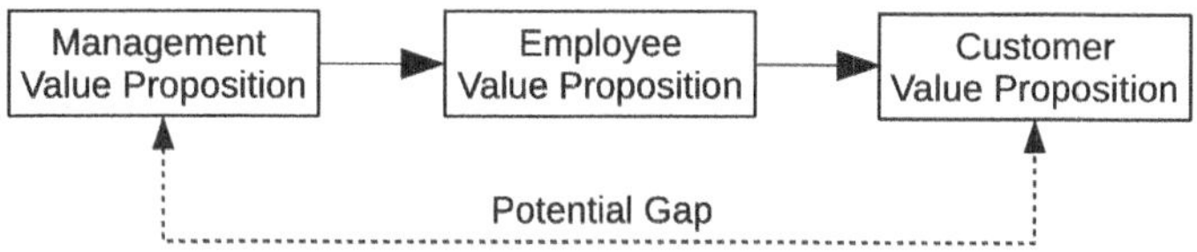

Figure 3.1 Value propositions in the value chain.

3.2 THE ANTHROPOLOGY OF WATER CONSUMPTION

The field of anthropology offers a unique insight into the consumption of tap water and its place within society. This information is invaluable for a water utility managers to better understand how they can add value to the community they serve.

Descriptions of domestic water consumption mostly revolve around physical variables, such as rainfall and consumption statistics. Water is described in cubic metres, milligrams, pascals and so on. While such descriptions of water are useful for a scientist or engineer, they generally are meaningless to consumers. Numerical descriptions are limited because they ignore the significance of the sociocultural dynamics of water consumption (Askew & McGuirk, 2004). To improve our understanding of the societal role of water utilities and their relationship with customers, tap water can be analysed through the lens of the cultural anthropology of water.

Moving beyond numerical descriptions of water, and appreciating the social, experiential and affective qualities of water consumption enables utility managers to improve their understanding of what it means to consume water as a practice of everyday life and the value that water represents in the lives of people (Allon, 2006). These insights assist in crafting meaningful water utility value propositions and help to develop stronger relationships with customers and the community.

Cultural anthropology is the field of social science mainly interested in how human beings interpret and add meaning to the world in which they live. Marketing and anthropology are well suited for intellectual and practical cooperation, as both disciplines place people at the nexus of enquiry. Anthropologists use their perspective to assist organisations in gaining insight into the lives of consumers and the behaviour of employees. Anthropologists involved in market research analyse the symbolic meaning of the creation, transaction and use of products. They use ethnography to enhance the understanding of consumption patterns and provide valuable insights for the marketer (Tian, 2013; Wayman, 2008; Winick, 1961).

The use of water has been of interest to anthropologists for a long time, initially limited to the role of water in extinct or non-European cultures. Modern scholarship in the anthropology of water is predominantly concerned with questions regarding the impact of privatisation, the right to water, and other socially critical issues (Wagner, 2013). Research into the role of urban water in the developed world is uncommon. The anthropology of contemporary attitudes and behaviours concerning water in developed societies is an emerging stream of research that discusses the cultural notions and values of reticulated water (Strang, 2004). Tap water is attracting the attention of anthropologists and other social scientists because of the way this essential service is embedded in daily life.

A methodological problem in water consumption research is that it is so common to everyday life that it retreats into the 'background of awareness' (Shove, 2003, p. 2). Water is essential to people's daily lives but it is so common that it only becomes visible once the supply is no longer available. Anthropology contributes to studying the everydayness of water through its focus on individuals instead of societal patterns. The anthropological view of water consumption enhances the technological view that dominates our industry.

Acknowledging the contributions of anthropology to our understanding of domestic water consumption helps water utility professionals to better articulate the value of water. Anthropology provides a language of water use that goes beyond the technological narrative widely used by water utilities. The value of these narratives lies in their outsider perspectives that can shift the way industry professionals view their activities to a more customer-centric point of view. Anthropologists study water consumption from two perspectives, that of the consumer and that of the water utility.

3.2.1 Water consumers

Even though water is an essential service, only a small percentage of the water consumed in developed countries is necessary for survival. The World Health Organisation (WHO) estimates that only about twenty litres per person per day is required for survival, with seventy litres per person per day to meet most personal needs (Reed & Reed, 2011). In contrast, the actual average water consumption in the developed world is double to quadruple the amount recommended by the WHO (Hoekstra, 2013). The minimum amount of water required to survive is even lower than suggested by the WHO. Recent research in the health sciences disputes the claim that people need to drink at least two litres of tap water per day as most of the required water intake is ingested through food and other fluids (Tsindos, 2012).

The limited need for water to survive diminishes the position of tap water as an essential service. It supports the idea that most of the water supplied by utilities is a discretionary purchase that shapes the lives of consumers beyond their physiological needs. The true value of water resides in its ability to shape the social lives of consumers. Water is used for cooking, bathing, swimming, recreation and functions as a symbol of purity in ritual performances, such as baptisms (Wagner, 2013). Water plays both a physical and a symbolic role in the lives of the customers of water utilities. The symbolic value of water is as essential to social life as is the physical value of water to biological life (Ten Bos, 2014). To understand the value proposition for water utilities, the symbolic dimension of water consumption needs to be untangled.

In pre-modern societies, water was both a life-giving substance as well as a potential destructive element. Water plays a primal role in creating mythologies, signifying its important place within daily life (Ten Bos, 2014; Wagner, 2013). These societies combined pragmatic water engineering with mythology and ritual practices to ensure their continued survival (Angelakis *et al.*, 2012).

Within Australian indigenous cultures, water is the organising force of *country*, which is the foundation of the Aboriginal cultural experience. The transformative power of water is embodied by the mythology of the Rainbow Serpent, a common motive in indigenous Australian cultures. The serpent rises from permanent water holes and travels across the sky, forming clouds and rain. The stories associated with this mythology discuss the relationship between humans and water, both as a means of survival and as a cultural value. These stories also have a practical value as the knowledge required to manage water resources is embedded within these stories (Barber & Jackson, 2011).

Balinese water temples, the *subaks*, are contemporary examples of how rituals conducted by priests, combined with pragmatic engineering, ensure a sustainable water supply for Balinese rural communities. The sustainability of the Balinese water system is not the sole result of rational engineering projects, but an emergent property of the interaction between the physical and symbolic aspects of water. The rituals organise the social reality of water users and the old lore contains pragmatic knowledge to maximise crop yields (Lansing, 1987; Lansing & Vet, 2012).

These two examples highlight the importance of the social life of water, both as a physical substance and as an intangible symbol, a life sustaining and transformative liquid that shapes social life (Strang, 2004; Ten Bos, 2014). The entanglement of the tangible and intangible benefits of water is not unique to traditional cultures and applies equally to developed nations enjoying clean and reliable tap water.

3.2.1.1 Indoor water consumption

Indoor water use is mostly a private activity, which takes place in closed spaces associated with privacy, pleasure and sensuality. Water users don't experience water as the consumption of units of H_2O, as it is expressed by utility professionals, but through the benefits they derive from the water. Water plays a utilitarian role in flushing toilets and washing dishes, a sensual role when sharing a bath with a loved one and acts as a catalyst for creative ideas when enjoying a hot shower. These benefits are predominantly intangible, which contrasts the numerical way in which water consumption is described by utilities (Allon, 2006; Allon & Sofoulis, 2006).

Many aspects of using water inside the home are a form of hedonic consumption. This type of consumption encompasses multi-sensory, fantasy and emotive aspects of product use and is related to the imaginative construction of reality by the consumer (Hirschman & Holbrook, 1982). Hedonic water consumption is a recent development. Over the past decades, the use of water in the bathroom has been transformed from a utilitarian to a hedonic type of consumption. For previous generations, it was uncommon to enjoy a daily bath or shower. During the weekly bathing process, the same water was usually shared by more than one person.

Contemporary life demands an ever increasing degree of comfort and cleanliness, which is expressed in the way water is used in the home (Allon, 2006; Allon & Sofoulis, 2006). Expectations about sweat and natural bodily odours have shifted from being a normal part of life to being socially unacceptable, leading to increased showering and laundry frequencies in areas where tap water is available (Sharp, 2006; Shove, 2003).

The contemporary indoor use of water is motivated by the Three Cs: *Comfort*, *Cleanliness* and *Convenience* (Shove, 2003, p. 396). Tap water facilitates the comfortable lives that contemporary citizens are accustomed to by providing reliable and safe supplies. It ensures the cleanliness of the body, of our clothes and of the home and it provides the convenience of home delivery without having to spend any time obtaining it.

The Three Cs of tap water are illustrated by the fact that in the twenty-first century, American households wash an average of 392 loads of 3.4 kg; more than twice the amount of laundry washed in 1950 (Shove, 2003). Water use is influenced by the interplay between the development of technologies and societal change. Social and technical changes are different aspects of the same phenomenon (Bijker, 2012; Bijker *et al.*, 2012). Shove (2003) describes this development as a ratchet, a mechanism consisting of teeth on the edge of a bar and a wheel that allows motion in only one direction. This analogy illustrates that once a technology is socially accepted, it is almost impossible to return to the previous levels of consumption. The current levels of comfort and cleanliness have developed in unison with developments in technology, such as the washing machine and the shower. These developments moved current consumption to historically high levels. There is as such a reciprocal relationship between the amount of water used, cultural values and the state of technological development.

Although the use of water has lost most of its religious connotations in the western cultural context, its use can in some occasions, nevertheless, be viewed as a profane ritual. The consumption of water is not a ritual in the common definition as a habitual action performed without understanding of its meaning, but as a transformative experience imbued with meaning beyond the practical purpose of the activities themselves (Habel *et al.*, 1993). A ritual is an expressive and symbolic activity that occurs in periodic and dramatically scripted sequences, performed with formality and inner intensity (Rook, 1985; ?). Consumer products, including tap water, are artefacts that assist and facilitate the ritual's objectives (Gentina *et al.*, 2012; Rook, 1985).

The daily grooming ritual, in which a hot shower plays a central role, is a personal ritual that transforms the private self into the public self. Having a bath can be viewed as a cleansing ritual during which the body is purified from the everyday world. The purpose of human grooming activities is not only to maintain hygiene but also to modify the body's colour, odour and perceived shape to match societal expectations. Grooming is an expression of body language, used to communicate an individual's social identity and water is a pivotal artefact in this private ritual.

Understanding consumer behaviour as ritual behaviour illuminates the psychological depth of everyday water use. Bathing satisfies three personal objectives: positioning the self in society, maintaining the body, and experiencing pleasure. The contemporary popularity of daily, or even twice daily, showering is held in place by strongly held beliefs of the ideal self, the therapeutic properties of hot water and a sense of pleasure (Shove, 2003). Indoor water use is thus to a large degree a form of hedonic consumption that assists people to shape their public identity.

3.2.1.2 Outdoor water consumption

Outdoor water consumption in Australian and American suburbs accounts for almost half of all the water being used in urban areas (Askew & McGuirk, 2004). Gardening cannot be undertaken without water and understanding the cultural aspects of this activity assists in understanding how water is consumed. Anthropological research on gardening reveals a positive relationship between people's emotional attitudes and feelings towards their garden and water consumption patterns. The stronger the emotional relationship, the higher the amount of water used (Allon & Sofoulis, 2006; Askew & McGuirk, 2004).

Whereas indoor use of water often occurs in the private seclusion of the toilet and the bathroom, outdoor water use occurs in the public space of the garden. The garden performs multiple social roles: providing a private haven away from the world of work and politics, functional space for leisure, a space contributing to a sense of home and place, and as a space for active engagement with nature (Bhatti & Church, 2004; Duruz, 1994). The strong attachment people have with their gardens can be understood against the changing patterns of domestic water consumption over the past decades. Homes and gardens have become important sites of leisure, recreation, and function as a public display of individual identity (Askew & McGuirk, 2004).

The importance of gardening to water consumers was demonstrated during the Millennium Drought that gripped large parts of Australia. At the height of this period, water utilities were forced to impose strict water restrictions, with some areas not allowed to use any water outside the house. These restrictions enabled the establishment of a cottage industry of water saving devices and promoted inventiveness on the side of consumers. Many consumers developed ingenious ways of recycling and saving water in the home for reuse in the garden. Others laboured carrying buckets of grey water from the bathtub or shower cubicle to the garden (Allon & Sofoulis, 2006).

Gardening can be understood as an accumulation of cultural capital and the expression of a social identity. Cultural capital refers to the non-financial assets accumulated by people to promote social mobility. Cultural capital is accumulated through immaterial means, such as education and socialisation, and through material means, such as clothes and consumer goods. During the process of accumulating social capital through gardening, water use is influenced by both forces of social distinction as well as social conformity (Askew & McGuirk, 2004). Social conformity is the force that compels people to shape their garden in accordance with social expectations. Social distinction is the force that drives people to shape their outdoor space as an expression of their identity; the garden becomes a means of self-actualisation.

The garden is objectified cultural capital and is strongly related to the lifestyle aspirations of the gardener. The garden, as a public space, is ideally suited to express individuality within the framework of conformity. The garden expresses individuality by creating a space that matches the lifestyle aspirations of the owner. Personal outdoor space is increasingly commodified, driven by the growth of the Do-It-Yourself industry (Bhatti & Church, 2004). This mechanism is juxtaposed with the fact that the garden is also a space for social conformity, with a well-manicured lawn as the symbol of belonging to the middle class. The garden thus plays an important display role in the accumulation of cultural capital, both to differentiate personal identity and a signal for social conformity (Askew & McGuirk, 2004).

3.2.2 Water utilities

The anthropology of domestic water use shows that the social and cultural dimensions of indoor and outdoor water consumption are vital phenomena for developing a nuanced understanding of domestic water consumption patterns. However, this cultural understanding of water has not been embraced by water utilities. The technocratic narrative of water utilities is largely myopic of the intangible benefits of water use sought by consumers (Allon, 2006; Allon & Sofoulis, 2006; Morgan & Smith, 2013).

Water management is predominantly a technological endeavour with a utilitarian view on water consumption. The supply of water is typified by engineering achievements (Angelakis *et al.*, 2012; Bell, 2012; Russell, 2009). This attitude created a 'dispersed system of domestic users who naturally came to expect, and eventually take for granted, the normalised presence of water in their households' (Allon, 2006, n.p.). The water consumer only notices tap water when it does not flow from the tap the same way it always has (Euzen, 2003).

Water utilities are predominately managed by technical professionals who view their work as the quantitative and utilitarian management of water. This is expressed in a propensity to emphasising water's tangible value over its intangible value, as is apparent in the label 'water utility'. This utilitarian attitude can be traced back to the dominance of engineering in water utilities and a preoccupation with the physical dimension of water (Morgan & Smith, 2013). Anthropologists found that water utility professionals are often unaware of the intangible value of the service they provide. They seem to be prone to ignore the symbolic dimension of water and have a limited understanding of the intrinsic value of water to the personal lives of the people they serve (Allon, 2006; Allon & Sofoulis, 2006; Morgan & Smith, 2013; Sharp, 2006). The narrative around water supply is dominated by stories of the 'heroic, monumental engineering efforts' to control the flow of water (Allon, 2006, n.p.). Large-scale water engineering is the ultimate expression of technology-driven modernity and unlimited rationality (?).

This myopic view of tap water as the result of technology transposes itself upon consumers through the public discourse of water utilities and their identity as a technological organisation. Consumers are no longer able to recognise the essence of water unless they are forcefully removed from the unlimited tap water supply, otherwise enjoyed in blissful unawareness (Leonard, 2013). The domestic water tap is a 'symbol of a rational, comfortable, modern life' (Allon, 2006, n.p.), but it renders the source of the water invisible to the consumer.

Technological and scientific paradigms are at the heart of utility management because of historical developments. Rapid population growth in the nineteenth century along with a lack of sanitation and clean water caused cholera epidemics in the main cities. The development of water utilities as we know them today coincided with the realisation that these epidemics were caused by microbial contamination of drinking water (Goubert, 1989; Hrudey & Hrudey, 2004; Hunter, 1990). The construction of sewers to remove wastewater and the provision of continuous water supply to households not only dramatically improved public health, but also facilitated the emergence of cultural expectations of cleanliness and the development of contemporary household technologies, such as showers and washing machines (Bell, 2012).

3.2.2.1 Tensions between public and private benefits

The provision of uninterrupted water to individuals and businesses forms the backbone of modern cities. Individual preferences are accounted for by ensuring that supply meets maximum demand at all times. The balancing of demand and supply is expressed in resource planning methods through the 'predict-and-provide' model of water infrastructure provision. This model requires engineers to expand water and sewerage systems continuously to meet urban demand, without necessarily considering the social dimensions of water consumption (Bell, 2012).

The primacy of technology in water utilities and the resulting high level of service and normalisation of tap water has prevented the development of strong relationships between the water utilities and their customers. The average amount of time customers deal directly with their water utility is less than ten minutes per year (Burns *et al.*, 2011). Water consumption usually only enters the mind of the customer when tap water smells, looks or tastes bad, and when the water bill arrives (Bell, 2012; Shove, 2003; Van Vliet *et al.*, 2005). However, reviewing the relationship between utilities and their customers in more detail reveals a layer of complexity.

Parr (2005) provides a historical account of the relationship between customers and water utilities by analysing a major incident. In May 2000, the water supply of the Canadian town of Walkerton was contaminated with a dangerous strain of *E. coli* bacteria. As a result of this contamination, a significant proportion of the five thousand inhabitants suffered from gastrointestinal infections and seven people lost their lives. Parr argues that water utilities need to provide both 'good water' and 'safe water', which expresses the tension between the private and the social good (Laing, 2003). Good water signifies a supply that satisfies the expectations of individual consumers, whereas safe water is water that meets the expectations of public health and engineering professionals.

Although water utilities are positioned as a public service providing a social good in the form of safe water, customers are billed for their individual usage of a private good in the form of good water. This tension between social and private benefits expresses itself in how consumers assess the service quality of their water supply. Consumers predominantly use somatic inferences to judge the safety of water. Good tasting water is understood by customers as identical to safe water. The taste of water is interpreted as an immediate and continuous sign of goodness. Assessments of the taste of food have developed through evolution and serve to protect people from ingesting poison (Breslin, 2013). The safety of water can, however, only be determined using scientific methods and is mostly invisible to the consumer.

Notwithstanding the weak relationship between the taste of water and its safety, the taste of water is nevertheless an important parameter water utilities need to control in order to build strong relationships with their customers. To ensure the safety of water, most water utilities add chlorine to disinfect it and ensure that public health is assured. This addition has the unintended consequence of negatively impacting the taste experienced by customers. Although chlorination is often necessary to ensure the public good with safe water, it is a 'counter-intuitive promise of safety' for customers and does not relate to good water (Parr, 2005, p. 258).

This tension has also been recognised in industry literature and is referred to as the 'water quality paradox', which holds that regulating the quality of water supplies can reduce the level of customer satisfaction (Kot *et al.*, 2011). Technical practices, such as adding chlorine to disinfect drinking water, can be perceived by consumers as an unwanted exposure to chemicals. Perceptions of water quality and the acceptance of the presence of chlorine are not simply a matter of personal preference, but are influenced by social, cultural and historical factors (Nagata *et al.*, 2011).

Water safety does not ensure consumer acceptation of its quality. The water quality paradox places limitations on the extent to which water utilities can meet the individual expectations of their customers and minimise risk to public health at the same time. This paradox can lead to dramatic consequences as one of the contributing causes of the Walkerton tragedy was that the treatment plant operator felt pressured by the community to lower the level of chlorine due to taste concerns, which resulted in the contamination with *E. coli* (Parr, 2005).

Another example of conflict between the private and the public good, and the relationship between water utilities and their customers, is the addition of fluoride to promote dental health. Although the health benefits of fluoride are beyond doubt (Gussy *et al.*, 2008), some segments of tap water customers oppose this practice. The water quality paradox in relation to fluoride lies in the fact that the distribution of fluoride to individual consumers is disconnected from their willingness to consume it as water services cannot be individually differentiated. The health benefits of fluoride conflict with the individual preferences of some consumers, which evokes the water quality paradox. Even though the social benefits of fluoridation are considered to be greater than the private concerns of unwanted ingestion, for consumers who take issue with fluoridation, the quality of service is diminished (Awofeso, 2012; Mendoza, 2011).

3.3 WATER UTILITY ADVOCATES AND REGULATORS

The anthropological research into water utilities and their consumers was corroborated through a series of interviews with representatives of Australian organisations that advocate on behalf of water utility customers and regulators of water utilities. The interviewees represent customers in broader advocacy matters or advocate on behalf of individual customers experiencing problems with their water utility. The interviewed organisations regulate or facilitate the relationship between utilities and their customers to mitigate the adverse effects of monopoly structure. These organisations were selected because of their dual perspective of both the experiences of customers and the behaviour of water utility employees.

The questions focused on the respondents' perceptions of what constitutes a high level of service from the customer's perspective. They also focused around the relationship between the behaviour of water utility employees and the perceived quality of the service. Questions concerned core services, that is water supply, and supplementary services such as billing and information provision. Within each of these, perceptions of quality and the impact of organisational behaviour on the customer experience were discussed. Two of the interviewees are regulators and three interviewees provide consumer advocacy. To test for theoretical saturation of the information, an additional perspective was obtained by interviewing a former water utility customer service professional. The following two sections summarise the views of these stakeholders with respect to employee attitudes, behaviour and the customer experience.

3.3.1 Employee attitudes and behaviour

Concerning the attitudes and behaviours of water utility employees, the most common issue is the ability of service staff to understand the needs of individual customers when a problem arises, there is a need to 'tease out what it is they are concerned about' instead of indiscriminately following scripts. Issues occur when, for example, staff don't use the language of the customer as 'terminology is often the problem'.

Interviewees identified a range of potential problems regarding the level of customer focus, specifically with regards to complaints handling. Some organisations have installed 'specialised complaint-handling teams', trained in dealing with sensitive issues. Some water utilities were able to provide a high level of service by also implementing 'specialised ... hardship teams to deal with people who are struggling with affordability'. These are specialised competencies that require a 'high degree of empathy'. Due to the importance of human interaction in supplementary services, one respondent identified that 'a culture of actually wanting to understand complaints, wanting to ... find the root cause and ... prevent [future] complaints' is a necessary condition. The level of service, especially with regards to supplementary services was perceived to be related to 'how empowered their staff are'.

A customer orientation was considered to be a 'cultural issue' about 'what water [utilities] think about their role ... if they think their role is to ... provide good quality services to customers ... they are more likely to ... listen to what they need and resolve a problem'. This culture was seen as an overall attitude towards service provision: 'If they see their business as an engineering business ... then they're less likely' to understand their customers. A well-performing service provider was considered to have a 'culture of actually wanting to understand' customers. Interviewees regularly emphasised the importance of empathy towards customer needs and concerns. Empathy was seen to be related to 'not necessarily judging whether they're right or wrong or who is right or wrong' and wanting to 'try to understand what's happening with the customer'.

The monopolistic nature of service provision was linked to a low level of customer focus as there is no 'threat of losing customers', which can make service providers 'pretty passive'. The monopolistic nature of the industry was linked to organisational behaviour as it creates a 'risk-averse industry'. Customer focus is not always intrinsically motivated and a recent industry-wide service failure showed that it 'had to be pushed from the outside', as action was only taken after public and political pressure.

A recurring topic in the interviews was the public image of the service provider and the level of trust customers have in the utilities' ability to meet their expectations. Public image is largely related to the status of utilities as a non-profit and government-owned service provider. Customers expect that the financial viability of water utilities is not a major concern because 'the government would ... support them'. In line with this reasoning, customers also expect to be 'treated fairly', independent of concerns over the financial viability of the utility.

Water utilities can also be perceived to have a high technological focus and 'see their business as an engineering business and it's all about just providing the water'. A technological focus was expressed differently in core and supplementary services: 'if there's a water burst ... water businesses are extremely responsive. When it's a customer service issue it might be less so'. There is a 'tendency for water businesses to have a technical approach ... because a huge majority of what they do in terms of delivery is the technical delivery of water'. One respondent provided some words of wisdom on how water utilities should be managed in the future: 'traditionally run by engineers, done a good job being technical ... economists and accountants come in ... first revolution ... cost discipline, pricing structure ... [next] revolution ... anthropologists coming into the industry ... who understand how people behave'. These words evocatively illustrate an evolution from an organisation focused on professional issues to an organisation that focuses on customers.

The attitude of employees towards customers and service provision was reported to be influenced by their type of education and the department in which they operate. For engineers, it was noted that 'customer service is [perceived as] an overlay over that which may not be core to the work of engineers'. Another respondent expressed the differences between customer service and engineering disciplines by focusing on the approach to solving problems: 'technology ... behaves in a way that's very predictable ... with great accuracy, whereas ... they tend to perhaps shy away from what might be seen as sort of softer things, like understanding how people feel about the service'.

All respondents hinted at the possibility of a disconnection between engineering and customer service departments as a possible source for customer dissatisfaction. Teams are segregated due to their specialised and divergent competencies. The organisational practicality of segregated teams with specialised competencies can, however, lead to a 'part of the organisation that is focused on wanting to understand the customer and another part of the organisation wanting to deliver infrastructure'.

3.3.2 Consumer experience

Interviewees predominantly interact with customers that have one or more negative experiences with their services provider through consumer advocacy and complaints handling. Customers contact advocacy organisations when the initial service recovery attempt of the water utility does not lead to a satisfactory resolution. The interviews are therefore heavily skewed towards instances of service failure and recovery. Customers are, in the words of one respondent: 'not ringing ... to congratulate you'.

Customers only contact water utilities when 'the water … tastes different [or] looks different to what it normally does'. The level of service quality is, due to the high level of technological sophistication of water networks in Australia, not a major concern. Water utilities 'don't get many complaints around the core services of water'. The expectations for core service delivery were succinctly expressed by one respondent: 'efficient delivery of water services to their homes at a potable quality at a fair and reasonable cost … [without] interruptions to their delivery' and 'clean, potable water that's safe for people to drink and to use for other domestic purposes'. Customers were perceived to be mostly concerned about those aspects of water directly knowable to them. Respondents noted that there is 'increasingly an expectation about the aesthetic water quality', both visually and concerning taste and odour.

Unlike core services, supplementary services require some form of direct interaction between the customer and the service provider. The most prevalent topics raised by the interviewed customer advocates were related to billing and dealing with people experiencing financial hardship. The most often raised subject was managing customers that struggle to pay their bills and the importance of 'the ways in which water companies deal with that'. Interviewees stressed the importance of individual service, especially in cases of service recovery. Service providers implementing best practice offer a range of payment options, including 'incentive schemes to help [customers] stay on track'. This is in contrast with the core service, where differentiation is not possible as all customers within a service area receive water from the same source.

Respondents noted that customers of water utilities have a low level of involvement with the service and are generally 'passive receivers of service'. Expectations are simple and customers 'want very basic things'. Although the level of involvement of customers is considered to be low and expectations are basic, customers have high expectations of the reliability of the service: 'people remember that one hour when a slug *[sic]* of dirty water went through after a repair'. Customers expect seamless service provision, 'people expect the service to be there and have no concerns unless the service is not there' and customers only contact the service provider when 'something unusual or different has occurred'.

The interviews revealed a tension between achieving regulatory requirements and customer perceptions of quality. Organisations with a high focus on regulation use this as justification for providing service solutions, instead of meeting the individual needs of the customer. Respondents acknowledged that regulatory requirements should be achieved as a minimum to ensure continuity and safety of supply. Regulation does, however, only provide a minimum standard and 'water businesses are allowed to do what they don't necessarily have to do'.

Interviewees acknowledged that there can be 'tensions between the amount of prescriptive regulation and the ability of the water businesses to be innovative'. Due to government ownership and the monopolistic nature of the service, the relationship between service providers and customers is enshrined in legislation and regulation.

Respondents confirmed that regulation can frustrate good service outcomes as the rules 'become convoluted … hard to both enforce and to follow, because they are so prescriptive'. Having a relationship-based approach is the opposite of a regulatory approach because 'if the relationship is working you never go to the contract'. These responses are interpreted as meaning that a good relationship with customers will prevent having to rely on the regulations to resolve disputes.

3.4 CRAFTING VALUE PROPOSITIONS

The anthropological literature and the interviews with consumer advocates and regulators show that the value of a tap water can be viewed from at least two perspectives: that of the supplier and that of the customer or consumer. The anthropological literature on water consumption emphasises intangible social and psychological values, which contrasts with the utilitarian approach taken by water utilities that emphasise the material benefits of water. Whereas, for example, the mundane activity of washing dishes is anthropologically viewed as a social activity undertaken within the family unit (Allon & Sofoulis, 2006), business scholars describe it as the volumetric consumption of a resource (Berkholz *et al.*, 2010, 2013; Fuss & Stamminger, 2012; Richter, 2010, 2011). This plurality of views doesn't imply that only one of these is correct.

The supplier's perspective of the quality of water is based on a scientific assessment of its chemical and biological composition, the volume supplied over time and the pressure at which it is supplied. The level of quality as perceived by customers is mostly based on trust in the utility and the aesthetic properties of the water (Elliott, 2010; Falahee & MacRae, 1995; Krishna & Morrin, 2008; Teillet *et al.*, 2010). While water utilities are mostly interested in providing *safe water*, customers expect to receive *good water* (Parr, 2005). Both the intangible and tangible aspects of water are necessary conditions for creating value but neither of these is by itself a sufficient condition of quality. Defining a value proposition can assist water utilities in merging these two perspectives in order to be more customer-centric.

3.4.1 Engineering and social science

The importance of the social dimension of water supply became apparent in a water resources development project in Sweden. The project had a long lead time and planning had been based on traditional deterministic forecasts. Shortly after completion of the project, the increase in water consumption unexpectedly ceased. The unanticipated reduction in water consumption was theorised to have several causes, such as lower birth rates, changes in development policies and higher water prices (Erlenkotter *et al.*, 1989). Social research into the drivers of water consumption and the social-economic environment could have mitigated the risk of over-engineering. This example shows how both social science and engineering are required to maximise value to the community.

Value is a subjective concept which implies that both the intangible and tangible views of water supply are equally valid within their own frame of reference. When crafting value propositions, the intangible values of water are dominant. When designing water systems to deliver this value, the numerical engineering approach is dominant.

In a customer-centric utility, both views of the value of water should be aligned as closely as possible. An effective value proposition for water utilities performance should thus incorporate the perspectives of both the customer and that of the utility, explaining *why* the service is provided and *what* the service consists of. The consumer's point of view justifies why water managers do what they do, while the scientific view of the water utility explains what needs to be done.

3.4.2 The value proposition canvas

After these theoretical considerations about the value of water, practitioners will want to know how to use these insights to craft a value proposition for their water utility. This section describes how this can be achieved using a Value Proposition Canvas. This canvas is a template used in workshops to collaboratively design the management value proposition. A canvas is essentially a large-scale template that can be recreated on whiteboards or butcher's paper to work on the problem in groups. The Value Proposition canvas is inspired on the business model canvas popularised by the lean startup movement (Osterwalder *et al.*, 2014). The canvas presented in this book is based on a revised version by blogger Peter Thomson (2013).

In practical terms, a value proposition is the place where your service (the what) intersects with the customer's needs and wants (the why). The water utility canvas includes the perspective of customers (the outline of a head) and that of the service provider (the glass of water). Each of these two perspectives is divided in three categories. For the customer, the utility needs to understand his or her perception of risk (fears), needs and wants. On the supplier's side the value proposition is defined by the customer's experience, the benefits they derive from the service and the features of the service that provide these benefits (Figure 3.2).

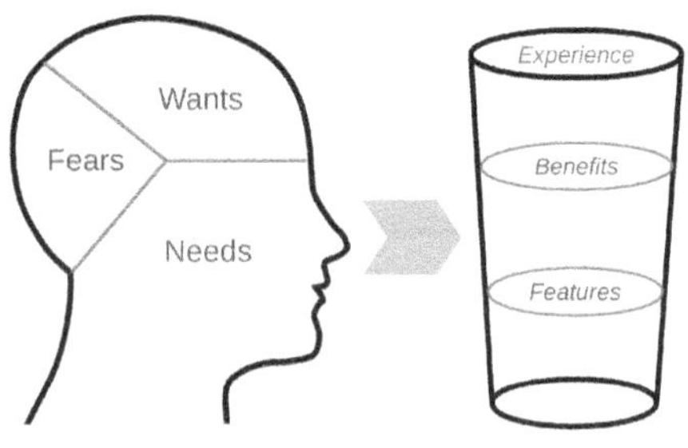

Figure 3.2 Value proposition model for water utilities.

This value proposition canvas is not a mathematical formula that will automatically lead to something useful as long as the correct inputs are provided. The main purpose of the canvas is to act as an incubator of ideas and to assist water managers to think about the services they provide beyond science and engineering.

3.4.3 The customer profile

The customer profile clarifies the customer's needs and wants, based on a psychological view of marketing. In this view, needs and wants differ from the common sense understanding of needs as essential and wants as optional, as explained earlier in this chapter. The psychological nature of this dimension of the value proposition requires the manager to empathise with customers. This empathy is based on listening to customer feedback obtained through engagement and research. The most important requirement of displaying empathy is to listen to customers instead of telling them what they need and want.

The customer's psychology forms the foundation of the value proposition. The value proposition is designed to meet the needs and wants of the customer. From this principle it follows that each market segment should be considered separately. Industrial customers will have different needs and wants than urban customers and even within the domestic customer segment, various smaller segments exist (Burns *et al.*, 2011). Customer segmentation is discussed in more detail in Chapter five. The different needs and wants of each segment implies that, in principle, each of these customer groups requires a different value proposition.

Within the customer's perspective of the value proposition there are three salient parameters: needs, wants and fears. Each of these three parameters is discussed in more detail below.

3.4.3.1 Needs

In its most basic form, a need is a customer's innate drive to make a purchase. We can use Abraham Maslow's (1943) *Hierarchy of Needs* to illuminate this aspect. Maslow studied positive human behaviour and analysed the lives of high-achieving people. Maslow's view of humanity is based on a humanistic psychology which identifies self-actualisation as the highest need for human beings. Physiological needs are most commonly associated with the consumption of water, one of the three essential substances. To craft a meaningful value proposition, however, it is essential to incorporate the higher order needs: safety, social belonging, self-esteem and self-actualisation.

Managers should empathise with the customer, looking at the world from their point of view without judging their needs and values. Identifying needs can be an abstract exercise and although needs are causally prior to wants, it is often best to define the wants before the needs.

When completing this aspect of the canvas, you need to ask yourself why customers need water. To this customer segment, is water mainly a utilitarian input to a process or a means to survive. What values do your consumers hold towards water conservation and water quality? Are gardening or other forms of outdoor entertainment important to your customers? Ideally, this section should not contain any references to water consumption itself but merely seek to describe the inner motivations that lead to the use of water.

3.4.3.2 *Wants*

The wants in a tap water value proposition consist of the different ways in which the relevant customer segment uses water. The obvious uses of water are basic activities such as drinking, cooking and personal hygiene. It is important to identify all types of water consumption for each segment based on their needs. Each need is met with a specific want for the service.

Although it is generally thought that water utilities have no competition, this is only the case when limiting their mission to providing tap water. However, ignoring competition and viewing water utilities as providers of tap water only would be considered myopic, as there are many alternative market offerings that provide the same benefits as tap water.

The canvas should include any market offerings that a customer can choose to use to meet their needs. For example, gardening is only one way to manage private open space and many alternatives are available to consumers. The alternative products are not necessarily competitors but should be considered to develop a well-rounded value proposition.

The main question to be asked when completing this section is how does a consumer meet their need through tap water. The alternative question is how this need can be met using alternative means.

3.4.3.3 *Fears*

Fears are the risks that customers associate with being a customer of the water utility. The word fear is preferred over the more common term risks because it better describes the psychological nature of the value proposition. Although water utilities are very adapt at managing risk, this term is too abstract for customers to comprehend. Focusing on fears instead of risks acknowledges the psychological importance of this dimension.

Various customer segments have different fears. If a community has a high number of people who are concerned with the use of chemicals in water treatment, then the value proposition needs to acknowledge this. In countries with regular droughts, the fear of water being rationed could play a role in crafting a value proposition. Fears are an equally important consideration for providing water services to commercial customers. Some businesses rely on water as an input into their production process and have very high requirements regarding reliability.

When completing this section, the question to be asked is what risks consumers perceive to exist when consuming this service. The concerns mentioned earlier about chemicals in tap water and fear of water restrictions are obvious. When considering, for example, digital metering, the fear of loss of privacy or the presumed negative health effects of electromagnetic radiation need to be taken into account. Whether these fears are rationally justifiable is irrelevant as the view of the customer is the final arbiter. The customer is always right about how they feel. A value proposition can play a role in understanding these fears and help the utility to reduce them.

3.4.4 The service

On the other side of the value proposition ledger we find the services provided by the utility. The features of a service describe *how* the service is provided. The benefits section describes *what* the service does for the customer. The customer experience describes what it *feels* like to use the service. This dimension should include both core and supplementary services: Every activity a water utility undertakes to meet the needs and wants of the customer. These activities are not limited to the harvesting, purification and supply of drinking water, but should also include facilitating aspects such as information provision, billing, payment facilities and so on.

3.4.4.1 Features

A feature is a factual description of the product or service. For the core service, this is the least important aspect of the market offering because its features are not differentiable. Water quality parameters are almost always regulated to protect public health. The aesthetic parameters are more flexible than the health parameters but potable water is not necessarily palatable water. The regulatory and discretionary parameters of water quality should be given equal importance. Water might be potable from a scientific point of view but that does mean it is palatable from the customer's perspective. A widget of water might be perfectly safe to drink from a scientific perspective, the moment of truth experienced by the customer can still be negative due to problems with water aesthetics.

For the supplementary services, the features are an important consideration. Supplementary services are what differentiates the water utility in comparison with other public service organisations and utilities. Some consideration are whether the service be provided face-to-face, via the internet or over the phone. Is the service fully automated or does it involve human interaction, and so on. Given that the supplementary services are the only time the customer directly interacts with the utility, these services are the perfect platform to fully express the value proposition.

3.4.4.2 Benefits

The benefits describe what the service does for the customer. The benefits are the ways in which the features of your service improve the customer's life by increasing their pleasure or decreasing their pain. The best way to list out the benefits is to imagine all the ways in which your services improves your customer's life. These benefits link directly to the needs identified in the first part of the value proposition canvas, such as a need for belonging or need for survival.

The Three Cs of tap water defined by anthropologist Veronica Strang (2004): Comfort, Cleanliness and Convenience can be used as a guide to describe the benefits of tap water in a general sense. For commercial customers these benefits are more related to being able to conduct their business as they see fit. Each segment might have slightly different benefits. In this section, it is also important to consider both core and supplementary services.

3.4.4.3 Experience

The customer experience consists of the collective moments of truth. The attitude of a customer towards the service provider is influenced by the accumulation of these experiences. The service experience is the core that describes the essence of the utility's brand.

The service experience can be best described as the way the costumer feels when consuming the service. The customer experience is bounded by the combination of features and benefits described above, but is also influenced by the needs and wants. The service experience is not simply the sum of the features and benefits but describes the emotional experience of the customer. Given the essential nature of our industry and the fact that tap water resides in the background of everyday life, these experiences will not be grandiose statements but subtle allusions to what customer do.

The customer experience is equally important in core and supplementary services. The difference between them is that core services are undifferentiated and cannot be used to brand the utility. In the core service of water supply, the customer experience is defined by the aesthetic properties and the pressure of the water and by the reliability of the service. In sanitation services, the customer experience is paradoxically defined by an absence of any experience.

The supplementary services are the way a water utility can differentiate itself from other service providers. What is the ease of doing business with the utility? Does the water utility communicate in a formal or informal manner? Does the customer feel welcome when lodging a complaint? These are the types of questions that need to be considered when completing this section.

3.4.5 Completing and implementing value propositions

The value proposition canvas should ideally be completed by a group of people in order to capture a broad range of views. Print or copy the canvas in Figure 3.2 on sheets of butcher's paper or reproduce it on a whiteboard to facilitate the discussion. In my experience, completing this exercise with water professionals and MBA students, many people struggle to move away from focusing on the features of the service instead of recognising the psychological aspects of the value their organisation provides. Many people who work in an organisation are used to viewing the service offering from the inside-out and live in a world of technical language, performance indicators and so on. The art of crafting value propositions is to view the organisation from the outside-in and use the language customers would use.

This chapter does not provide a completed example of value propositions because they should be based on the collective knowledge within each organisation and not on generic statements. The canvas should also not be perceived as a rigid model that has to be completed to the letter. A value proposition matching the needs, wants and fears of the customer should be related to the benefits, features and experience of the service provided by the water utility. Any model that achieves this principle will assist in developing a meaningful value proposition.

The objective of completing the value proposition canvas is not necessarily to leave the room with a beautifully crafted statement. The objective of the canvas is to create a narrative of the value proposition for each segment. The purpose of this narrative is to inform any decision making with respect to the marketing mix.

Although in many instances the value proposition for tap water might seem obvious and generic, there will be meaningful differences between water utilities and customer segments. The purpose of crafting the value proposition is to develop a common language of who the utility's customer segments are and how they can be best served. The canvas helps to structure the discussion around the value proposition and guides the development of a meaningful document. Value proposition canvases are valuable tools to position the service provider in the mind of the customer.

The management value proposition developed in this chapter is the starting point of the value chain that determines how further activities are shaped. For example, acknowledging bottled water as a competitor leads to water utilities thinking about activities that promote tap water. Recognising that some communities are less inclined to accept chemicals in tap water might shape the way communication is targeted, and so on. The ultimate sign of success is when customers perceive the value proposition in the same way as management has originally crafted it. Customer-centric service provision occurs when the perception the customer has of the service matches the design of the service, when the *why* and the *what* of the value proposition come together. The completed canvas assists management to develop their internal marketing, to define service quality and to cultivate strong relationships with customers. Each of these three aspects are discussed in the following chapters.

Chapter 4

Internal marketing

About twenty years ago, a realisation developed that the disciplines of management and marketing needed to be integrated in order to achieve superior service (Daniel, 1998). This realisation led to the development of internal marketing which is a 'philosophy of valuing and treating employees as an intermediate set of customers inside the firm and enhancing the value provided to employees with the aim of encouraging them to enact the organisation's marketing objectives' (Lings & Greenley, 2010).

If marketing is interpreted as the process of meeting the needs and wants of customers and providing positive experiences, then internal marketing can be best described as the process of meeting the needs and wants of employees and delivering positive experiences for them. Internal marketing is the best way to help employees to create an emotional connection to the service they provide, beyond their technical involvement (Mitchell, 2002). The purpose of internal marketing is to minimise the gap between the customer value proposition and the employee value proposition, shown in Figure 3.1. Internal marketing aligns with the axioms of Service-Dominant Logic (S-D Logic) as value is co-created between the customer and the organisation. For co-creation to occur, employees need to understand the needs and wants of the consumers. They need to empathise with the experience that the customer has when consuming the service.

The marketing triangle shown in Figure 4.1 visualises the relationship between the various types of marketing activities. The relationship between management and customers sets the value proposition and is referred to as external marketing. This requires management to understand the value that customers seek from the

doi: 10.2166/9781780408675_077

organisation, as described in Chapter three. Employees deliver the promise through interactive marketing, and their direct and indirect relationships with customers. Internal marketing closes the loop and involves management activities that enable customers to deliver the promise to customers.

Figure 4.1 Marketing triangle. *Source*: Unknown.

Peter Drucker (1954, p. 37, emphasis added) described marketing as the activity that informs 'the *engineer*, the designer and the manufacturing man [*sic*]' about what customers want. Drucker's description of marketing identifies some of the actors in the value-creation process, such as engineers and manufacturers. To co-create the value proposition, the needs and wants of customers are communicated through the value chain from designers and engineers to the people who deliver the service. Successful marketing requires effective communication and internal coordination between business functions (interfunctional coordination), through the whole of the value chain, both within and outside of the firm. Successful marketing requires the designer, the operator and the marketer to work in close unison.

The previous chapters showed that, although water utilities are service factories where customers and employees rarely interact, the behaviour of employees indirectly impacts the customer experience. The interviews summarised in Chapter three provided an external view of this mechanism and this chapter further investigates these findings by analysing the attitudes and behaviours of water utility employees from an internal perspective, with particular reference to the role of engineers.

This chapter discusses internal marketing as it relates to water utilities. The first section describes some of the common issues related to customer focus in water utilities. The second section discusses professional orientation, which is a situation where employees are more focused on their professional task than on satisfying the needs of customers, and therefore create a potential barrier to customer focus. The following section discusses the relationship between engineers an customer-facing employees and how this can lead to negative outcomes for customers. The penultimate section discusses these two concepts within the context of water utilities. This chapter closes with some reflections on how to practically implement internal marketing in a water utility.

4.1 CUSTOMER FOCUS IN WATER UTILITIES

The interviews with various stakeholders in the water utility value-creation network described in Chapter three provided some insights into the relationship between management and marketing. The interviews revealed the relationship between the attitudes and behaviour of employees and the experiences of customers, summarised in Figure 4.2.

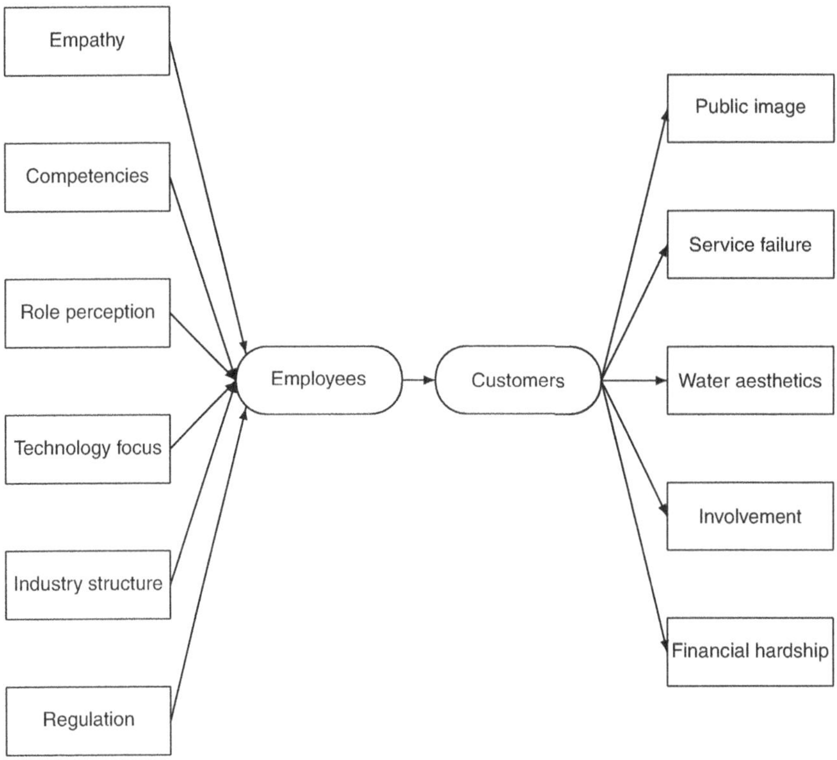

Figure 4.2 Customer experience salient parameters.

For employees, empathy with customers and their role perception were found to be salient topics with respect to the experience of customers. Interviewees reported that, if the role perception of utility employees is based on technology, then the understanding they have of the needs of customers can suffer. The respondents believed that the type of training and competencies of staff (engineering versus other types of education) are an important indicator of their orientation towards customers. Besides these personal characteristics, employees are also influenced by the monopolistic industry structure and the type of regulation within a water utility's jurisdiction.

With respect to customers, the public image of the water utility was believed to influence their perceptions of quality. Due to the high reliability of water services, and the essential nature of the service, any service failure is amplified. Water being a physical product, the visual aesthetic quality of the water was found to be a major determinant of service perceptions. The level of involvement of consumers of tap water was considered to be low, except when the service fails. Due to the essential nature of tap water, respondents considered managing customers that suffer financial hardship to be a core task for water utilities.

The exploratory research presented in the previous two chapters suggests that water utility management is dominated by the physical sciences, which is a contributing cause of service failures. The anthropology of water consumption describes how the technocratic narrative of water utilities is largely myopic of the intangible benefits of water use sought by consumers. In other words, water utilities often communicate the value of water by referring to its essential nature from a biological perspective and by highlighting their technological achievements, rather then the social benefits that water consumption provides.

The positive result of this focus on technology is efficient and reliable pressurised drinking water service directly into the houses of consumers. Tap water has become the norm for civilised life, residing in the background of daily life. The negative impact of the great reliability and normalisation of water is that societies are disconnected from the natural resource from which water is extracted and that managers of water supply systems are disconnected from consumers (Allon, 2006; Allon & Sofoulis, 2006; Bell, 2012; Morgan & Smith, 2013). The focus on technology required to deliver the high levels of expected reliability is a likely cause for some of the criticism regarding customer orientation raised at water utilities (Ahmed, 2009; Auriol & Picard, 2009; Deichmann & Lall, 2007; Karbowiak, 2003; Skellett, 1995).

Accounts of significant service failures, discussed in Chapter two, are examples of the disconnection between the behaviours of employees and the needs of customers (MacGillivray, 2014; McCoy, 2014; Parr, 2005). Analysis of such incidents reveals a tension between the social and private benefits provided by water utilities. In general terms, the social benefits of a service rely on the professional judgement of the employees, while individual consumers can only assess the private benefits, as shown in Figure 1.2. Where professional judgement and personal preferences of consumers don't overlap, the risk of service failures increases.

An example of this tension is the addition of fluoride and chlorine to tap water. Both chemicals are beneficial to public health, but some consumers view these chemicals as unwanted additives (Awofeso, 2012; Kot *et al.*, 2011; Mendoza, 2011).

This tension is inherent to this industry because the necessarily undifferentiated nature of tap water prevents matching individual preferences. For example, people who object to the taste of the water or the presence of certain chemicals cannot be provided with water that matches these preferences. In core water services, social benefits necessarily prevail over the private preferences of consumers because minimisation of public health risk is the prime directive of water utilities.

The literature provides a clear example of disconnection between customer perception and professional judgement of water quality. In the *Asia Pacific Public Relations Journal*, Elliott (2010, p. 75) refers to public discourse about water quality, after commissioning a desalination plant in Sydney, as 'absurd' and 'nonsensical'. These strong words indicate that Elliott is unable to empathise with the perception of customers, prioritising the scientific view of water quality over the subjective experience of consumers. Anecdotally, this type of tension is prevalent within water utilities. In personal communication with a water utility professional, this sentiment towards customer preferences was expressed poignantly (private LinkedIn correspondence 2013, used with permission):

> I don't see the point of focusing so heavily on making utilities customer focused. You have to deliver good quality services but I don't get the concept that the customer is best placed to decide what those services are. They know they need pressure and they know if they don't have water out of their tap, but beyond that, they have only a small understanding of the what is required to run a safe and efficient water supply system.

Interviews with customer advocates and regulators, described in the previous chapter, identified a *professional orientation* of employees as one of the impediments of customer-centric service provision. The attitude of employees towards customers and service provision was reported to be influenced by their type of education and the department in which they operate. For engineers, it was noted that 'customer service is an overlay over that which may not be core to the work of engineers'. One interviewee suggested that differences between the objective nature of technology and the subjectivity of customer perception could lead to tensions in the relationship between utilities and their customers.

The relationship between engineers and marketers has been extensively discussed in marketing literature, predominantly within the context of new product development in high-technology organisations. Conflict between marketers and engineers, the so-called *engineering–marketing interface*, in the technology sector is seen as a pervasive and as-yet unresolved problem (Keaveney, 2008, p. 653). The causes and consequences of conflict between marketers and engineers have been extensively researched in new product development for the high technology industry, but have not been investigated within the context of water utilities.

4.2 PROFESSIONAL ORIENTATION

The interviews with customer advocates, discussed in Chapter three, showed how the attitudes of utility employees moderate their behaviour and subsequently impact customer outcomes. The attitude that was most often observed by respondents as potentially preventing beneficial customer outcomes can be best described as a *professional orientation*. This orientation is the antithesis of a customer orientation and implies that the employee is more focused on internal processes than on the individual needs of the customer. The interviews showed that a professional orientation has three dimensions:

- Mangers prioritising financial performance over assisting customers experiencing hardship;
- Prioritising procedures and regulation over individual service and;
- Solving engineering problems instead of solving customer problems.

A professional orientation is closely related to a tension between professional judgement and individual customer preferences (Bastian, 1998; Laing, 2003). Public services are characterised by a difficulty to satisfy individual needs because the benefits are primarily societal in nature. Services, such as defence, public health and environmental protection, are motivated by social benefits and collective consumption rather than satisfying individual preferences. In these types of services, the judgement of the producer of the service, or of the professional making decisions, is dominant.

Tap water is a public service where individual users are able to be identified as each house is separately connected to the water system. This direct relationship with individual users renders the judgement of consumers dominant. Water utilities operate within a tension between public and private benefits. The professional orientation construct describes in three dimensions the extent to which the attitude of water utility employees prioritises public over private benefits and professional judgement over consumer judgement.

4.2.1 Financial hardship

Interviewees emphasised the importance of a water utility's ability to provide special services for customers experiencing financial hardship. This is a unique competence for utilities and other essential service providers as consumers have no choice but to purchase their services. Their responsibility to minimise risk to public health risk requires water utilities to ensure the supply of safe water to all members of society, including those that cannot afford it.

Assisting customers who suffer financial hardship is not explicitly discussed in marketing literature. Most research into this issue has been published in sociology literature, which places financial hardship within the context of adverse life events. Financial hardship is viewed as the result of events that lead to an accumulation of disadvantage, related to life changes such as employment, marriage and health

(Maroto, 2015). Experiencing financial hardship is thus a profoundly personal issue to the people who suffer from it.

With respect to financial hardship and the provision of essential services, regulators and customer advocacy organisations have commissioned research reports on this topic (CUAC, 2013; Hall Partners & Open Mind, 2011). These reports emphasise the psychological and sociological aspects of hardship and the specialised competencies required to satisfy the needs of this customer segment. In response to this increased attention, utilities have implemented special considerations for customers suffering financial hardship, such as specially trained teams and flexible payment plans (Skellett, 1995).

To provide services to customers experiencing financial hardship, employees must be able to understand and empathise with their personal circumstances. Employees with a high level of professional orientation are considered to lack this empathy and prioritise corporate financial outcomes over concerns for the customer.

4.2.2 Service bureaucracy

The second aspect of professional orientation is prioritising standardisation and regulation over meeting the personal needs of customers. Reticulated drinking water supply is regulated to provide internal motivation for the industry to ensure a certain degree of customer focus is assured in a monopolistic environment (Byatt, 2013; Kot *et al.*, 2011). Regulations exists in relation to the technical requirements for service provision and in some instances requirements for supplementary services are also formally defined (ISO 24510:2007).

Processes and procedures are a necessary part of providing service to large numbers of customers. Service blueprinting, contact centre scripts, journey mapping, and so on are valuable tools to ensure customers are assisted efficiently and effectively (Lovelock *et al.*, 2007). There is, however, a risk of establishing a customer oriented bureaucracy, which results in tension between the individuation and differentiation required in contact with customers, and the coordination and standardisation requirements of work (Korczynski, 2004). A balance between the need for uniformity of core services and individual customer service is essential in managing customer oriented bureaucracies (Tomar & Dhiman, 2012).

The tension between standardisation and individual service is of particular interest to the water industry because of the tension discussed earlier between public and private benefits. Providing social benefits requires adherence to regulations to minimise risk to public health through the addition of chemicals to the water. Strictly following regulations can in some instances lead to a conflict between social benefits and individual preferences, as illustrated by the water quality paradox.

While differentiation of core services is not possible in tap water, supplementary services can be tailored to individual needs. Service work tends to be underpinned by a binary logic: standardisation and customer-orientation. These two streams of thought are in potential contradiction to each other. A high level of professional orientation is a situation where employees prioritise standardisation over personalising services.

4.2.3 Technical orientation

The last aspect of the professional orientation discussed by interviewees was the prioritising of technical matters over social concerns. The importance of the technical dimensions of the delivery of core services can lead managers to prioritise these aspects over individual customer concerns. This can escalate into conflict between technical compliance and the customer-focused ethic of the marketing concept (Morgan, 1990; Rexha *et al.*, 2000).

A technical orientation is the antithesis of customer orientation and comprises an attitude held by employees and management that prioritises the technological dimensions of service provision over needs and wants of customers. A high level of professional orientation in this dimension implies that technology is an independent objective instead of a boundary condition and is instrumental to providing benefits outside of engineering itself (Trevelyan, 2014).

The research on technical orientation is dispersed and no agreed definition or scale exists to measure this phenomenon. The existence of a technical orientation was first discussed in a study on healthcare management (Hunt *et al.*, 1975). The attitude of managers in healthcare management sets the parameters under which employees conduct their activities. The technical orientation of higher level managers negatively influences coordination among employees of different functions (Hunt *et al.*, 1975; Loh *et al.*, 1995). A high level of technical orientation by managers has also been found to engender a perception that the organisation they work for is a less progressive place that emphasises process over accomplishments (Loh *et al.*, 1995). A study in engineering services in Australia found that managers with stronger technical values exhibit weaker market-oriented behaviours (Rexha *et al.*, 2000).

Within the field of organisational psychology, professionals with a high level of technical orientation were found to be less receptive of personal praise. In the computer industry, a technical orientation was found to be negatively associated with communication skills. Communication style varies between different levels of technical orientation (DuBrin, 2005). Klein *et al.* (2002) recommend that, for project teams, a mix of orientations is desirable to achieve the best outcomes: a strong technical orientation, empathy with users, and organisational awareness should all form part of the ideal team.

The last two dimensions of the professional orientation overlap to some extent. In tap water, a professional orientation expresses itself as the idea that customers ought to be satisfied when the water meets regulatory standards. The social benefit of tap water is invisible to the consumer as it is characterised by the absence of disease. The provision of social benefits is a credence quality that requires trust from the consumer. Social benefits are dominated by professional judgement, which requires a technical orientation. This technical orientation is enshrined in rules and regulations, which is indicative of a service bureaucracy.

4.3 ENGINEERING–MARKETING INTERFACE

The engineering–marketing interface is an expression of interfunctional coordination, which is the coordinated utilisation of resources in creating value for customers. In organisations with a high level of interfunctional coordination, every individual contributes fully to the creation of value for customers (Narver & Slater, 1990). A high level of interfunctional coordination implies that all actors in the value-creation network co-produce in an environment free of conflict. A high potential for conflict between professional groups indicates that coordination between functions is weak, lowering the level of interfunctional coordination and thus potential to create value.

The literature on the interfunctional coordination between actors in value creation networks is mostly abstract in the sense that employees and managers performing different functions are considered to be uniform groups. These general considerations are of limited use in managerial practice because the specific roles of employees influence the organisational dynamics. To thoroughly understand the value creation process, research needs to be relevant to the specific roles of people in the value creation process, rather than context-free managerial research (Jaworski, 2011).

Organisational theory is based on a division of labour that separates employees into specialised functional groups to most effectively perform the required activities. Although division of labour is efficient, it is prone to develop into interfunctional conflict (the often lamented silo mentality) when dealing with tasks that require a high degree of coordination between teams, such as new product development (du Fossé, 2010; Keaveney, 2008).

The division of labour also separates classes of employees from the customers they service. The second foundational premise of S-D Logic states that indirect exchange masks the fundamental basis of exchange. This premise implies that the division of labour and the development of complex vertical marketing systems masks the fact that all activities are geared towards the co-creation of value with the final beneficiary being the customer. Most employees of water utilities are working backstage of the service performance, invisible to the customer due to the low requirement for direct interaction.

Before the industrial revolution division of labour was minimal and marketing and engineering were integrated. When somebody wanted a table, she would speak directly to the carpenter who translated the needs of the customer into the table she needed (Vargo & Lusch, 2008). The ongoing division of labour requires an organisation to implement internal marketing initiatives to ensure a high level of interfunctional coordination and focus on the needs of the customer.

4.3.1 Engineers and marketers

Two specialised roles essential in providing services based on technology are engineering and marketing. Engineers provide the technical expertise to innovate and develop technology and marketers connect this process to customers.

The extant research on the role of engineers in the marketing process shows that a strong relationship between engineering and marketing sections promotes the successful development of new technological products (Keaveney, 2008; Lancaster, 1995; Shaw & Shaw, 1998, 2003; Workman, 1993). Conflict in the relationship between engineers and marketers obstructs the successful development of new technological products, resulting in lower product quality (Gupta *et al.*, 1985; Keaveney, 2008; Song & Thieme, 2006).

The importance of healthy relationships and the existence of conflict between marketing and engineering personnel has also been acknowledged in engineering literature (Taylor, 2011). Engineers are highly specialised professionals that need to turn their attention to customer issues to ensure products and services match the needs of customers. This literature acknowledges that 'for an engineer to have a successful career, it must contain an element of organisational behaviour and marketing' (Edwards, 1995, p. 2).

To empower engineers to contribute to the marketing process, they need a broader understanding. This broadening of engineering skills is most commonly achieved through professional development and education, often in the form of an MBA (Edwards, 1995; Hoske, 2013; Norman, 1997; Taylor, 2011; Visser, 1996; Workman, 1995).

Early literature on this topic found that marketing and sales played less a central role than engineering for product development in high-technology firms. New products were developed to be functional and cost-efficient based on the assumption that customers would be satisfied in this way (Workman, 1995, 1993). This foundational research inspired further investigation into the relationship between engineers that design and develop products (R&D), and the professionals that sell those products, the marketers.

In the wake of this research, the product development process in high technology firms has moved from an engineer-dominated environment to a marketer-dominated situation (Keaveney, 2008). The design of personal computers, which evolved from functional beige boxes to aesthetically designed functional devices, is an example of how marketing and engineering can work together.

4.3.2 Sources of conflict

Previous research in the technology sector identified several contributing causes of potential conflict between engineers and marketers. Differences in tasks, goals and objectives and stereotyping of personality traits of the two professional groups have been identified as causes (Shaw *et al.*, 2004; Shaw & Shaw, 2003). Certain types of conflict between these two professional groups can act as a barrier to effective cooperation (Keaveney, 2008; Shaw *et al.*, 2004). Conflict between marketers and engineers can also be positive because it not only defines a relationship but it also helps to stimulate genuine interest in preserving the relationship (Keaveney, 2008). Conflict about cognitive issues enhances decision quality while affective conflict reduces the quality of decisions (Langfred, 2004). In other words, conflicts about the issue can lead to improvements, while conflict between people reduces the quality of the final product.

Management literature identifies three types of interfunctional conflict. *Task conflict* is an 'awareness of differences in viewpoints and opinions about a group task' (Jehn & Mannix, 2001, p. 238). This type of conflict is functional because it stimulates divergent thinking and can lead to improved decision making. Avoiding task conflict can hinder a group's ability to develop creative solutions (Keaveney, 2008). *Process conflict* is an 'awareness of controversies about aspects of how task accomplishment will proceed' (Jehn & Mannix, 2001, p. 238). Process conflict reduces productivity, increases role ambiguity and interferes with the quality of the task (Keaveney, 2008). The third type of interfunctional conflict is internally directed *relationship conflict*, which is the focus of this chapter. This kind of conflict, 'awareness of interpersonal incompatibilities, includes affective components such as feeling tension and friction' (Jehn & Mannix, 2001, p. 238). Relationship conflict is dysfunctional, based on personality and difficult to resolve. Relationship conflict negatively impacts group performance, employee satisfaction, cohesiveness and mutual understanding (Jehn & Mannix, 2001).

Engineering is a highly specialised profession that has regularly been criticised for not understanding its role in the value chain, resulting in substantial research into the relationship between the functions and characteristics of engineering and marketing (Keaveney, 2008; Shaw *et al.*, 2004). Engineers and marketers have different educational backgrounds and work under incommensurable paradigms. Engineers consist of a single professional group with a formalised education, based on the physical sciences while marketers stem from more diverse backgrounds and originate from the social sciences (Shaw & Shaw, 2003).

Qualitative research by Keaveney (2008) identified a range of stereotypes held by the two groups about each other. Engineers perceived marketers to be arrogant and prone to exaggerate. They perceived marketers to be too emotional, not sufficiently concerned about details and lacking understanding of technology. Negative assessments were symmetrical between groups as marketers saw engineers as being too rigid, lacking creativity and having unsuitable social skills. They also found engineers to be poorly customer-focused and lacking understanding of business requirements.

These findings replicate earlier research on the topic (Gupta *et al.*, 1985; Lancaster, 1995; Shaw & Shaw, 1998, 2003). Following attribution theory, these stereotypes are considered to cause conflict between the two groups. This theory holds that when conflict occurs, either party will attribute a different cause to the conflict, which in most cases is based on personal variables. The attitude that engineers hold towards marketers as professionals, and vice versa, serves as an internal justification for the observed behaviour and are a major source of relationship conflict (Keaveney, 2008).

More recent research in technology firms found minimal differences between the two groups and engineers were found to take a positive view of marketing and marketers, although stereotypes of marketers existed among engineers. The internal views held by engineers were found to be related to their type of education. Engineers who completed marketing training were found to be more likely to have a deeper understanding of marketing issues and displayed a more positive attitude towards it. However, their view of the relationship between the two groups was found to be the same as those not trained in marketing (Shaw & Shaw, 2003).

Sources of conflict are more likely to be grounded in the differences between these two professional groups than between their commonalities. The main difference between the two professions is their paradigmatic foundation. Engineering is based on the physical sciences and demands precision while marketing is a social science where certainty is not achievable. These paradigmatic differences lead to misunderstandings of language and methods for solving problems (Edwards, 1995; Griffin & Hauser, 1992; Shaw & Shaw, 2003; Sofoulis, 2010).

4.3.3 Engineers

The word engineer is ambiguous as it can refer to a mechanic, a train driver or a degree-qualified professional. Not all technical staff associated with product development are necessarily degree qualified. Engineering as an independent profession is a highly specialised endeavour, with each engineer trained for a specific discipline, such as civil or chemical engineering. In many countries, the title engineer is protected by legislation (Trevelyan, 2014).

The marketing literature regularly identifies differences in education as the root cause of conflict between engineers and marketers. Since the early research into attitudes of engineers towards marketing, much has changed in the way engineers are trained. Non-engineering skills such as multidisciplinary teamwork, communicating with people from different backgrounds and ethical conduct are common learning outcomes for graduate engineers in contemporary engineering curricula (Grasso & Burkins, 2010; Gribble *et al.*, 2006; Trevelyan, 2014; Workman, 1995).

For engineers to have a successful career, their activities must contain an element of management. A broad understanding of business disciplines, including marketing and organisational behaviour, are essential for an 'engineer's survival in the corporate jungle' (Edwards, 1995, p. 1). Not everybody agrees with this assessment as an earlier quoted textbook for water utility engineers sees marketing as a 'peripheral activity' (Brown, 2010b, p. x). To mitigate the risk of over-specialisation, it is common for engineers to undertake postgraduate management training, usually in the form of an MBA, which is seen as a mechanism to enhance their so-called soft-skills (Jaiswal, 2014; Mintzberg, 2005). There is still a sense that the inclusion of soft skills in the engineering skill set is lacking (Selinger, 2004). Resistance to including social science could be related to the use of the terms soft and hard skills, as this betrays a negative attitude within the engineering community towards 'soft' knowledge that is not based on the 'hard' physical sciences.

4.3.3.1 Engineering ethics

The tension between social and private benefits discussed previously can be understood when considering the ethical theory of engineering. Public works engineering is dominated by a utilitarian logic that promotes the greatest good for the greatest number of people, where the benefits of the collective prevail over the individual. The forward-looking emphasis of utilitarianism on consequences and its desire to maximise collective benefits aligns with the aspirations of engineers, especially in public services. Methods of civil engineering generally prioritise the well-being of the whole of the community (the *civilis*) over the preferences of individuals. In water utility engineering, the greater good of public health is prioritised over individual concerns regarding the addition of chemicals to the drinking water supply.

Techniques such as risk analysis and cost–benefit analysis are based on the utilitarian method of hedonistic calculus, which seeks solutions that maximise social benefit. The utilitarian approach has unwanted consequences from a marketing perspective as it can preference the rights of the collective over the individual. The defence of the utilitarian to this problem is that utilitarianism is an anti-paternalistic moral theory that does not claim to know what is good for each individual but requires engineers to seek the view of all people impacted by their decisions. This ideal is, however, not practically achievable and maximisation of benefits for the collective will almost certainly lead to unmet needs or even damage to certain individuals (Carper, 1991; Nelson & Peterson, 1982). Service failures usually occur where engineering fails to seek feedback from those impacted by their decisions, highlighting the importance of market research in engineering projects.

4.3.4 Marketers

Viewing marketing as the process of creating value implies that all actors in the value creation network are part of the marketing team, including engineers. Marketing is a whole-of-business process that is not undertaken by a specialised team but by everybody in the organisation. Within the context of this book, a marketer is considered to be anyone whose primary role brings them in direct contact with customers, or somebody who is responsible for a function that communicates directly with customers. The term marketer is not common within the reticulated water industry. Organisations have differing names for the functions undertaken by people with specialised marketing competencies, such as customer service, customer support, communications, and so on.

Technical and customer functions in water utilities are divided across the provision of core services, which are based on engineering, and the delivery of supplementary services or customer service activities, which form part of the marketing scope of works. Although functionally different roles, the tasks of engineers and marketers often overlap.

This diversity is caused by the fact that marketing is methodologically more diverse than engineering. In marketing, both qualitative and quantitative methods are used to collect data and, both interpretive and positivist paradigms are used to draw conclusions (Lee & Lings, 2008). In contrast, engineering is exclusively a quantitative positivist endeavour. This methodological variety in marketing could explain the view of marketers that engineers are too rigid and lacking creativity (Keaveney, 2008).

The typical marketing curriculum is not based on a foundational science such as physics. The typical marketing curriculum includes various aspects of management and communication theory, information systems, economics, business statistics, business law, accounting and finance, as well as a range of specific marketing units. The skill set of marketers also includes a large creative component to design products and services and to create effective communication. Effective marketing practice involves a broad range of knowledge of the principles of sociology and psychology (Walker *et al.*, 1998).

4.4 ENGINEERS AND MARKETERS IN WATER UTILITIES

The engineering–marketing interface is not only an important parameter for organisations selling high-tech products but also for service factories in general and water utilities specifically. Engineers play a pivotal role in service factories, given the reliance of these service providers on technology to realise value propositions. Engineers don't only work in the development of new products but they are often also involved with operational aspects of providing service through technology (Schmenner, 1986; Thomas, 1978).

The literature on the relationship between engineers and marketers in new high-tech product development might not seem applicable to water utilities at first glance. New product development also plays a central role in water supply to service growing cities with tap water and to meet the demands of evolving consumer expectations (Erlenkotter *et al.*, 1989; Macomber, 2013; Reid *et al.*, 2010).

An important consideration for utility engineers is balancing the supply and demand for tap water. This balance can be achieved by either increasing supply through technological solutions or reducing the demand through social marketing. Balancing supply and demand through technological solutions is not primarily an engineering problem, but also a marketing problem because it involves predicting how consumers use water. Interpreting the balancing of supply and demand as an engineering problem can lead to inefficient investment in infrastructure (Erlenkotter *et al.*, 1989). Understanding water consumption requires a high level of sensitivity with respect to consumer lifestyle as the use of water pervades the details of everyday life (Allon & Sofoulis, 2006). Planning for sustainable water supplies requires a strong relationship between engineers and marketers to balance technology with consumer preferences.

In areas without a reliable natural water source, utilities can increase supply with alternative sources, such as recycled sewage. Consumer preferences play an even more important role when water utilities consider such sources. Public support or rejection of alternative water sources is influenced by the perception that people have of such sources of water (Dolničar & Schäfer, 2009; Dolničar *et al.*, 2014; Kemp *et al.*, 2012). The emotive nature of the use of sewage to supply drinking water requires sustained and frequent community information campaigns to counter opposing campaigns (Kemp *et al.*, 2012). When proposing to implement alternative sources, engineers and marketers need to work together to manage the introduction of alternative water sources.

When consumers distrust the social benefits provided by tap water, they are prone to seek alternative sources, such as rainwater tanks, bottled water, or personal filtration devices (Akpinar & Gul, 2014; Burlingame & Mackey, 2007; Doria, 2006; McGuire, 1995; Webber *et al.*, 2015). The use of chemicals to protect public health can be contentious issues which have the risk of eroding public confidence in tap water (Kot *et al.*, 2011; Mendoza, 2011). Public perceptions of tap water are very closely related to its aesthetic appeal, most importantly its flavour. The common dictionary definition of water refers to it as a tasteless substance. The absence of any taste and odour is the expectation that customers have of tap water. The majority of taste testing methods in tap water are scientific in approach and seek to identify compounds that cause flavours (Dietrich *et al.*, 2003). The taste of water is, however, caused by more than just its chemical composition (Krishna & Morrin, 2008). The taste of water is a complex subject that requires the expertise of engineers to identify the compounds that cause issues and the expertise of marketers to analyse customer perceptions. Neither of these two perspectives is sufficient to fully assess service quality.

The final activity where engineering and marketing interact in the management of water utilities is the development and construction of large infrastructure projects. The construction of such works impacts the amenity of the public space. To mitigate the potential impact, project managers involve the affected community (Croke & Williams, 2011). This involvement is primarily a marketing function that does not only involve customers and consumers, but any member of the community impacted by infrastructure works.

4.4.1 Organisational attitudes

The National Water Commission in Australia commissioned research to elicit the current views of water planners and managers about the social dimensions of urban water management and sustainability (Sofoulis, 2010). This research indicates that the industry has transitioned from a purely technical orientation to an environmental one and is moving towards a social focus. The interviewed managers were found to have a strong connection with the social aspects of water management.

The transition from an engineering focus to an environmental perspective, towards full inclusion of the social dimensions of water management poses a challenge to water utilities. The emphasis on averages and the mythical figure of the 'average water user' hides the diversity of society. The undifferentiated nature of the supply of tap water prevents utilities from realising the benefits of customer segmentation, commonly used by other types of service providers.

Sofoulis (2010) found that water managers view the current focus on social aspects of water management at variance with their primary training. The incompatibility of the forms of knowledge between the physical and the social sciences emphasises the need for translation between the two disciplines. The physical science-oriented water managers struggle with the interpretation of social science based research (Meissner, 2015; Sofoulis, 2010).

Sofoulis closed with a reflection on the unexpected strong connection of interviewees with the social sciences. The sample of respondents was possibly biased towards managers willing to speak to a cultural researcher and thus more likely to be positive about the value of social research in urban water management. However, conversational hints about resistance to this reorientation from within their organisations or professions indicate a gap concerning the integration of the social sciences in water utility management (Sofoulis, 2010).

4.4.2 Empirical research

The professional orientation concept and the engineering–marketing interface were empirically tested by surveying professionals from various water utilities. To obtain this information, 150 water utilities in the United States, South Africa, New Zealand, the United Kingdom and Australia were invited to participate in the research project. The response rate to the invitations was very low, with only seven formal consents received. Four utilities in Australia, one in the United States, one in New Zealand and one in South Africa formally agreed to participate. Data was, however, only collected from the four Australian utilities because the remainder did not proceed with data collection, without formally withdrawing.

Although no formal justifications were provided for the low participation rate, some managers informally reported that providing the requested information was seen as a high risk to their organisation's reputation, even though confidentiality was guaranteed. One manager expressed concern that the information could be linked to their organisation as the industry is small. He expressed that participation could lead to negative consequences due to the political sensitivity of the requested information. The reluctance to participate could be related to the high level of regulation of this industry. Surveys commonly form part of the regulatory framework and can influence pricing decisions.

A total of 357 employees submitted to the questionnaires between May 2013 and May 2015. The survey consisted of quantitative questions using seven-point Likert scales. Respondents were requested to provide additional qualitative information to explain their answers. The data was not collected to benchmark the participating utilities but to investigate the existence of a professional orientation and to understand the engineering–marketing interface.

4.4.2.1 Respondent characteristics

The majority of respondents work in a technical function, with other respondents associating themselves with administration and customer related functions. Technical functions include operational tasks, not all of which require tertiary qualifications. Only one of the employees in a customer function reported having a degree in engineering. About 14% of employees with engineering qualifications also reported having completed a tertiary degree in marketing, such as an MBA. The ratio of engineers to marketers among respondents is approximately two to one (Figure 4.3).

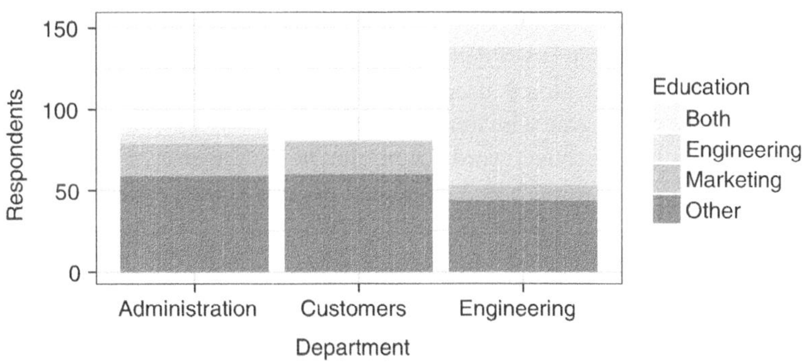

Figure 4.3 Employee respondent characteristics.

4.4.2.2 Professional orientation

The professional orientation instrument was developed to assess the extent to which employees are focused on professional issues, rather than on individual customer issues. The instrument items are based on the results of interviews with customer advocacy groups. The first four items relate to the financial dimension, the second four items to the procedural dimensions and the last items to the technical dimension:

p1 Customers that cannot pay their bill have the wrong financial priorities.
p2 Financial management is more important than the needs of individual customers that cannot afford water bills.
p3 Customers who are not able to pay their bill do not need our help.
p4 Customers who do not pay their bills on time should not receive service.
p5 The more detailed our rules and regulations are, the better the level of service we will provide.
p6 When we provide service in accordance with rules and regulations, customers will always be satisfied.
p7 Decisions must be based on rules and regulations instead of individual customer needs.
p8 Without detailed rules and regulations we are not able to provide satisfactory service to customers.
p9 The more we invest in infrastructure maintenance and development, the more satisfied our customers will be.
p10 Providing services is predominantly a technological activity, not a customer support activity.
p11 Customers do not have the knowledge to independently assess the level of service we provide.
p12 Providing services in accordance with technical standards and regulations is sufficient to meet customer needs.

This scale was deliberately worded in a negative way, assuming that a strong level of tension exists. This method is used to avoid an acquiescence bias, which is the tendency to agree or disagree with items, regardless of their content. The qualitative data shows that the negative wording of the items ensured that respondents provided thoughtful answers. Some expressed dissatisfaction with this approach. One respondent suggested that this negative wording could 'cause a poor workplace culture'. This comment reveals that this respondent assumed that the items are normative and supported by the researcher, which indicates acquiescence bias.

Comments from research participants related to customers suffering financial hardship were equally divided between those affirming and those contradicting the scale items. The word compassion in relation to customers facing financial hardship was mentioned three times. One respondent referred to providing assistance to financially struggling customers as a 'duty to help'. On the other side of the issue were those who expressed a lower level of empathy for customers who struggle to pay. One respondent went as far as suggesting that 'some customers are not able to afford our bills because they waste money on other things'.

The items related to the tension between rules and individual service provision attracted most comments. Respondents recognised the tension between needing to comply with regulation and providing individual service. Both positive and negative sentiments were expressed. An underlying theme in many of the comments was that 'providing service in accordance with said rule and regulations *should* provide customer satisfaction' (emphasis added). This comment expresses a strong belief in the efficacy of rules, which was also expressed by others, for example: 'decisions must be based on rules and regulations instead of individual customer needs'. The strong belief in rules was expressed poignantly by two respondents: 'too much individuality in customer service can and will lead to kaos [*sic*] in the long term' and 'there will always be 'nutcase' customers who will not accept the processes that are to be followed'.

Respondents also recognised that bureaucracy comes at a cost and that 'a lot of time, money and effort goes into meeting detailed rules and regulations ... resources that could be better spent on maintenance, innovation and customer service'. The predominance of providing service through regulation also creates tensions as it 'does not allow any flexibility for the water authority to help the customer's individual needs' and 'no matter how detailed the regulations and rules are you will never satisfy your customers'.

Several respondents mentioned that regulators should also be considered to be customers. One respondent suggested that scores would be different if the questions were about regulators instead of customers and another respondent suggested that 'more time and energy [is] devoted to satisfying auditors than to customer service'.

Although it could be argued that, from a traditional customer-centric perspective, regulators should only be viewed as a boundary condition, using the value creation network principles from S-D Logic and the principle of balanced centricity, regulators are beneficiaries and thus a type of customer. Within water utilities, regulators are the guardians of the social benefits provided by water utilities and in some cases also represent customers and regulate private benefits.

Traces of a technical orientation were found within the comments relating to engineering: 'the term 'customer satisfaction' does not to my knowledge appear in the project plans'. Respondents viewed the influence of politics as an impediment to a utility's orientation on customers as 'political direction ... will always override ... internal strategies and initiatives'.

4.4.2.3 Engineering–marketing interface

In the literature on this topic, a range of methodologies and instruments have been used. The engineering–marketing interface instrument consisted of eight seven-point Likert items, adopted from previous research (Gupta & Wilemon, 1991; Gupta *et al.*, 1985; Parry & Song, 1993; Shaw *et al.*, 2004). The original statements were amended as the term marketing is not commonly used in this industry and was replaced by the term customer service:

e1 Customer service takes a short-term view and engineering a long-term view.
e2 It is difficult for engineers to communicate effectively with customer service staff.

e3 Engineers and customer service staff cannot trust each other.
e4 Engineers and customer service staff have different values.
e5 Engineers and customer service staff do not understand each other.
e6 Friction between engineers and customer service staff is healthy.
e7 All engineers should know something about customer service.
e8 Engineers are more important to a water utility than customer service staff.

Comments on the relationship between engineers and customer service staff confirmed the importance of a strong relationship: 'the work engineers do is essentially a core customer service'. The majority of comments were normative: 'we *should* have the view that we all … achieve the best outcome for our customers' and 'there *should* be no engineering and customer service divide' (emphasis added). This confirms that the responses relate to an attitude of the employee and not their perception of the actual behaviour within their organisation.

Several comments confirmed the existence of tensions between the professional groups. One respondent noted that in their view 'engineers don't seem to take into account customer impacts'. Another respondent posed a thought experiment to express their lack of trust: 'would love to see engineers working in a customer service area [and] see if the water utility is able to function'.

The fact that engineers are generally removed from customers was also highlighted: 'while engineers are very important to a water utility, their work isn't always favourably received by customers. The involvement of customer support and communications staff in the engagement of customers is definitely needed.'

Some respondents provided insight into the cause of possible tensions. The backstage nature of engineering reduces opportunities to engage directly with customers: 'engineers are very project driven … and don't have to deal with the general public'. Another respondent emphasised that this is not the case as engineers regularly interact with customers. Two respondents emphasised personality differences, rather than professional training as a cause of conflict.

Comments provided by respondents confirmed that tension can exist between engineers and marketers, but also provided examples of positive cooperation. Since the original research on this topic, engineering curricula have been modified to include more subjects based on social science which helps engineers to better function in a multidisciplinary environment. This study confirms the findings of Shaw and Shaw (2003), which reported limited tensions between the two groups, but nevertheless found that some stereotypes still exist.

4.4.3 Conclusions

The quantitative analysis indicated that a professional orientation exists, with engineers showing a higher score on this construct than their colleagues in other professions. This can be explained by the procedural nature of engineering and its reliance on a utilitarian ethic that seeks to maximise benefit to the collective, even though it might disadvantage the individual.

The Engineering–Marketing Interface construct was used to identify differences in attitudes between engineers and marketers. Comments provided by respondents confirmed that tension can exist between engineers and marketers, but also provided examples of positive cooperation. Since the original research on this topic, engineering curricula have been modified to include more subjects based on social science which helps engineers to better function in a multidisciplinary environment. This study confirms the findings of Shaw and Shaw (2003), which reported limited tensions between the two groups, but nevertheless found that some stereotypes still exist.

The tension between social and private benefits is inherent to the work of water utilities. Core services are not differentiable and are generally provided by engineers. Marketers have a more direct relationship with customers through supplementary services. The tension between social and private benefits in tap water services constrains the extent to which consumer preferences can be integrated with new product development.

The attitudes of engineers towards the marketing of services are equally important as they are in the marketing of products. In services that rely on equipment, engineering plays an instrumental role in the delivery of services (Thomas, 1978). The behaviour of engineers directly impacts the customer's experience due to the inseparability of service production and consumption. Service factories are generally highly dependent on engineers because of the high degree of automation. Following the theatrical metaphor introduced by Grove *et al.* (2000), engineers in such industries are mainly employed in backstage roles and significantly influence how core services are delivered to customers, with front stage customer service staff managing supplementary services, such as complaints-handling or processing billing.

These results indicate that, although engineers have a higher level of professional orientation, they perceive the potential for conflict between marketers and themselves to be lower than other groups within the organisation. The higher level of professional orientation indicates that engineers focus more strongly on social benefits than individual benefits. The engineering–marketing interface construct does not measure the actual level of conflict between teams, but the perception that individual members have of the level of conflict.

4.5 IMPLEMENTING INTERNAL MARKETING

To be successful, management must first 'sell' the value proposition to employees before it can sell its services to customers. Internal marketing activities are closely related to any other types of management interventions to influence organisational culture. Given that water utilities are service factories where the majority of employees work backstage from the process of service provision, internal marketing is a productive way to ensure that all business processes are designed to maximise value for customers.

A range of activities can be deployed by management to achieve this objective. Customer service training is one of the most obvious ways to implement internal marketing. Besides equipping employees with theoretical knowledge, practical experiences are a powerful way to improve interfunctional coordination between the various teams. Developing customer personas, discussed in Chapter five, is another method to engender empathy with customers.

Job rotation can help to engender empathy in employees who otherwise are not exposed to direct interaction with customers. Job rotation also helps customer-facing staff to better understand the needs of their colleagues that deliver the service in the back office. One study reported on a job rotation program implemented by the Lake Victoria North Water Services Board in Kenya. After ongoing reports about negative customer experiences the organisation decided to use job rotation to improve organisational performance. Research by Tarus (2014) suggests that job rotation, as an internal marketing strategy, had a significant effect on performance in the workplace. The paper did not elaborate on the type of job rotations that formed part of this project. The researchers reported limitations to the effectiveness of job rotation because empowerment opportunities may be limited when employees perform routine production or service jobs. This approach was found to be more effective for jobs that have complex characteristics and a degree of independence.

4.5.1 Using taste testing as internal marketing

One of the most direct experiences that consumer have with tap water is its taste and odour, collectively known as flavour. Managing taste is a challenging proposition due to the complexity of water chemistry and the subjectivity of the human experience. Of the 225 IWA research papers discussed in Chapter two, 43 articles discussed managing the taste experience for customers. The importance of the taste of water as an expression of service quality is discussed in further detail in the next chapter. Employees that have control over the taste of water quality have little interaction with the customers that consume their services. Water utilities commonly use flavour profile analysis by a group of specially trained staff to test for aesthetic issues. Although this method provides scientific intelligence about the taste of water, it does not replicate the customer experience.

Replicating the customer experience for employees in the back office is a powerful way to create empathy for the needs and wants of the customer. A water utility in regional Australia has introduced taste testing by employees as a method to gather intelligence about the taste of water and at the same time recreate the customer experience in the office and develop empathy (Goetz & Prevos, 2016).

This event is called a Tap Crawl, which is a tongue-in-cheek name to internally promote the event. Employees of the utility are asked to taste a sample of water from each of the nineteen water systems they manage. The test is based on hedonic measurement using a seven-point Likert scale that records the extent to which subjects like or dislike the taste of water. Employees are also asked to describe the taste using the water flavour wheel (Suffet *et al.*, 1999). This method is more subjective than flavour profile analysis and requires a large sample of approximately thirty to fifty untrained subjects to provide a reliable result. The advantage is that hedonic testing approximates the customer experience as closely as possible and provides actionable intelligence for water treatment engineers and operators.

More important than providing interesting data, the Tap Crawl events have spawned lively discussions about the widely differing taste experiences caused by the variety of water sources. A most interesting result was that, although the laboratory testing and complaints from customers already indicated that the experience of customers in some systems was not optimal, it took the Tap Crawl event to bring this issue to the attention of all levels within the organisation, including the board of directors.

This internal marketing example has changed the attitudes of the employees of this utility towards the subjective customer experience. The results of the ongoing taste testing are now used in capital investment planning and the taste testing has been extended to also include customers. The taste of water has become a central topic in the utility's narrative about the customer experience and has played a central role in their most recent pricing submission to the economic regulator.

Chapter 5

Customer relationships

Customers of organisations that supply a basic service are not considered to be interested in a relationship their service provider. Practitioner experience and references in literature suggest that customers of water utilities have a low level of involvement with their service and have a weak relationship with their service provider.

Water utilities can benefit from developing a stronger relationship with their customers as it will improve the process of setting prices and achieves positive results in demand management (Karbowiak, 2003). Strong relationships between customers and their service provider are also beneficial after service failures because customers with a positive image of the utility are more likely to be accepting of unavoidable emergencies.

Relationships between customers and water utilities are often based on service failures as customers usually only contact the service provider when expectations have not been met. This fact suggests that direct contact between the utility and the customer should be minimised (Skellett, 1995). Reducing the amount of contact requires the prevention of service to increase the level of service quality. This mechanism is the basic idea behind the invisible water utility concept. A lower level of contact between water utilities and customers implies a higher level of service quality. The fact that strong relationships are beneficial as well suggest a paradox.

This chapter discusses various aspects of the relationship between water utilities and customers. The first section discusses how customers can be defined and understood. This section proposes a broad definition of what constitutes a customer and the value creation network. The first section also discusses methods to segment

doi: 10.2166/9781780408675_101

customers and how to use segmentation to strategically shape the service offering. The first section closes with a discussion of consumer involvement and shows that the level of consumer involvement in tap water services is, contrary to common belief, very high.

The remainder of the chapter focuses on the relationship between the utility and paying customers. The second section discusses three topics related to managing the customer experience. Customer complaints are often viewed punitively. Marketing research shows that they should be viewed as a gift to the organisation. Process mapping is a collection of techniques that can be used to improve the relationship between customers and the utility. Water utilities use promotion mainly to shape the demand for water. In most promotional activities, utilities position themselves as technological service providers, ignoring the intangible value of water consumption, which impacts the way consumer view their service provider. The last section of this chapter analyses the concept of the invisible water utility and discusses how the invisibility paradox can be resolved.

5.1 UNDERSTANDING CUSTOMERS

The first step of any marketing process is finding out who the customers of your organisation are and what their needs and wants are. Although identifying customers might be obvious for a water utility, some deeper reflection on this topic shows that this question requires some consideration. This section discusses the concept of value creation networks which shows that water utilities not only create value for urban water users but for many other stakeholders in the process.

5.1.1 Value-creation networks

In traditional marketing theory, value is delivered through a 'channel', which is not an asset to transport water but a set of interdependent organisations that manage the value moved through the supply chain. A marketing channel is traditionally viewed as the logistics of the supply chain for the delivery of products. Current scholarship about marketing channels includes all relationships between the service provider and any other organisation or person it relies on to deliver services to its end users.

Activities undertaken in marketing channels are characterised by a division of labour, either internally or between organisations. Each step in the value creation process contributes to the value proposition through its own competencies, being either production, advertising, customer relationship management and so on. At the macro level, marketing channels describe the relationships between participating organisations, while at the micro level, individual employees of the service provider are considered. Traditionally marketing channels were viewed only at the macro level but this concept can be extended to include individual employees or groups of employees within channel partners (Svensson, 2005).

Most scholars view marketing channels as linear relationships between manufacturers, suppliers, wholesalers and so on. The reality of business is, however, more complex than this simplified model can describe. Linear marketing channels simplify the complexity of marketing and risk losing sight of other relationships. To understand the complexity inherent in marketing channels, Svensson (2005) proposed the concept of *spherical marketing*, which recognises that marketing channels do not have a clear beginning or end and are circular in nature. Information travels down the marketing channel from the service provider to the consumer, but also from the consumer back to the service provider. Value is not created in a linear process from the service provider to the consumer, but co-created between both. The spherical marketing concept requires a holistic view of the actors, activities and resources, beyond the presumed start and end of traditional linear marketing channels.

Service is not created in a one-way relationship between the supplier and the customer. Instead, value is created through a network of actors (Gummesson, 2008). The third axiom of Service-Dominant Logic (S-D Logic) states that all social and economic actors participate in creating value, and the fifth axiom places the service process within its societal context (Vargo & Lusch, 2016). These axioms provide a macro-perspective of marketing channels that clarifies relationships between the various channel partners, including the employees of the firms. S-D Logic replaces the common B2C (Business-to-Consumer), B2B (Business-to-Business) and C2C (Consumer-to-Consumer) approaches to marketing with the generic Actor-to-Actor (A2A) model.

The marketing channel is a value creation network where all actors co-create value (Lusch & Vargo, 2014). All actors in the marketing channel share a common goal, which is value creation through resource integration (Vargo & Lusch, 2004). In the A2A approach, both the producer, the consumer and every person or organisation that assists in delivering a value proposition are part of the marketing channel. The A2A approach expands traditional value chain models to a network of actors that collectively produce the value proposition. The network can be studied at the organisational level, or at the micro level of individual interaction. Value creation networks include employees of the service provider, employees of channel intermediaries and customers.

The A2A approach is based on earlier work by Bagozzi (1974, p. 78) who describes the process of economic exchange as a set of 'social actors, their relationships to each other, and the endogenous and exogenous variables affecting their behaviour'. A social actor can be any individual who contributes to the value creation process. Endogenous and exogenous variables are causal concepts that influence the relationship between actors, which can include social, psychological or physical factors.

Within the context of this study, the attitudes of water utility employees are examples of causal concepts. The psychological constructs and social structures of employees influence the organisation's value proposition. Chapter four discussed how the attitudes of employees in water utilities and their internal relationships can impact the value provided to customers.

Value creation networks can be visualised and analysed using graph theory, which was also used in Chapter two to define the water utility marketing mix. Relationships between actors in the network are indicated with directed lines. The direction of the lines indicates the beneficiaries of the exchange as value flows towards the beneficiary.

5.1.1.1 Value-creation networks for water utilities

These principles can be used to describe the value creation network for water utilities. The simplified network in Figure 5.1 visualises the distance between engineers and customers and the disconnection of the customer from the environment, which was highlighted in the anthropological literature discussed in Chapter three (Allon, 2006; Bell, 2012). Engineers have a direct relationship with the environment through their task of extracting, purifying and supplying water. Engineers and scientists generally only have direct contact with internal staff that interact with customer (the marketers) and with regulators. They are mostly isolated from direct contact with the end user of the service they provide.

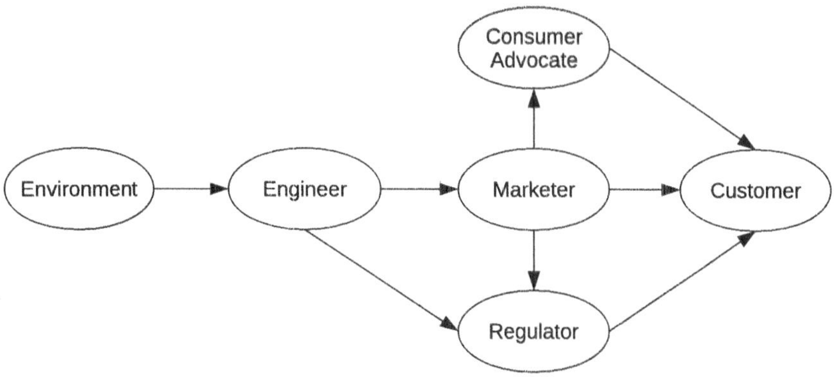

Figure 5.1 Simplified value creation network for water utilities.

Regulation mitigates the monopolistic nature of the industry by ensuring that water utilities act in the best interest of the community. Regulators are included in the value creation network through their role in assuring service standards and advocating for customers to prevent monopolistic behaviour. Both customer advocates and regulators are positioned between the utility and their customers as they communicate with both parties.

This network also demonstrates the concept of balanced centricity, which is a situation where a service provider maximises value for all actors in the network. This concept can be extended to not only recognise human actors but also the natural environment. Within the network sketched in Figure 5.1, the natural environment is recognised as an actor in the value creation process. Water utilities are fully integrated with the natural environment through their reliance on the natural hydrologic cycle (United Nations, 2012). This integration into the natural water cycle requires utilities to be cognisant of their role within the water cycle.

The equivalence of the natural environment and other human actors also has practical implications. The natural environment requires a share of the total amount of water in a catchment, which will not be available for human consumption. Government agencies can act on behalf of the environment and purchase and sell water as required. If the environment is not considered an equal actor, all water would be available to be acquired by rights holders, depleting the natural environment of water. Recognising the environment as an actor in water markets prevents a tragedy of the commons, a situation where the environment is an externality, depleted through self-interest of market participants (Garrick *et al.*, 2012; Hardin, 1968).

Using value creation networks is a powerful way to visualise and analyse the interdependencies within an industry. The relationship between engineers and employees dealing directly with customers was discussed in Chapter four. The remainder of this chapter focuses on the relationship between the water utility and the consumers of its services.

5.1.2 Customer labelling

The word customer is a limited term that indicates a commercial relationship between an individual and an organisation for the purpose of economic exchange. Using this label is restrictive because a customer is not necessarily the person consuming the service. This difference is expressed in the distinction between customers and consumers. However, as public service providers, the responsibilities of water utilities reach beyond their contractual relationship with customers and include the community as a whole.

The international standard for the assessment and improvement of water services (ISO 24510:2007) describes a customer as a 'registered user'. Water utilities generally keep records of who their consumers are through their billing database or customer relationship management system. Most such systems equate customers with a connection to the reticulation. Each connection has one or more registered users.

These two paragraphs show that various labels for customers exist depending on the context in which we define them. In the narrow view, customers are those people which can be identified by the water utility as being responsible for paying the bill; in the broadest context the utility looks after the whole of the community.

The IWA journal articles discussed in Chapter two contain a range of synonyms to describe the beneficiaries of water utility services. These labels are interchangeable as some articles use customer and consumer in the same context (Franceys, 2006; McGuire, 1995; Mugabi & Njiru, 2006). This interchangeability is an artefact of the monopoly conditions in which utilities operate. Within any given service area, all members of the community consume water, the majority of households are connected to the water system and within each household there is at least one customer. There are significant semantic differences between these terms, but the interchangeability of their usage indicates that these labels refer to the same people.

The labels used to describe beneficiaries of services provide insight into the relationship between customers and their service provider (Plangger *et al.*, 2013; Syed *et al.*, 2011). Drawing from sociological labelling theory, words (labels) used to denote customers influence the nature of the relationship between service providers and customers. The terminology a firm uses to label its customers alters how customers expect to be treated in their dealings with them. For example, a student has a different relationship with his or her service provider than a client or a patient. Using the word student or patient implies an imbalance in power where the customer is not always right. Using the word customer signifies a relationship that is weighted the other way.

Labelling theory suggests that nouns used to refer to people are mental shortcuts that extend beyond their dictionary entry. A label activates a cognition about a group of individuals, which affects how members of the organisation think about and act toward that group (Plangger *et al.*, 2013). For example, the word 'consumer' can be interpreted as more distant and impersonal than using 'customer'. A consumer is an anonymous person who consumes, while a customer has a relationship with the service provider (Syed *et al.*, 2011). The label ascribed to somebody defines the relationship between the person being described and the person attaching the label.

Within the context of marketing many labels are used to describe customers: 'passenger', 'guest', 'patron', 'shopper' and so on. The label relates to the expectations customers have of the level of service and the perceived obligations of the service provider. Labels define the psychological contract between the service provider and the customer, which can be visualised with a relationship ladder. More personal terms, such as 'partner', score high on the obligation and expectation dimensions, while labels such as 'user' indicate an impersonal relationship with low expectations for both parties (Plangger *et al.*, 2013).

Customer labelling is also recognised within the water utility marketing literature. A case study on the development of a customer orientation for Sydney Water discusses the 'realisation that the board no longer dealt with 'ratepayers' but with customers, and that customers were real people' (Harrison & Stamp, 1991). A case study on Wessex Water discussed their 'programme to change the culture … and stop thinking of 'consumers' and start thinking of 'customers' … ' (Skellett, 1995, p. 24). A reverse logic has also been proposed. Allan *et al.* (2013) argued that poor peri-urban people in developing countries should not be labelled as 'consumers', paying market rates for their water, but as 'citizens', with a right to access clean water. The citizen label relates to the identity of water utilities as public service providers because privately owned companies have individual customers and public services provide for the whole of the community (Franceys, 2006).

5.1.2.1 Customer labelling in water utilities

Evidence of customer labelling theory can be found in the use of customer labels in the IWA database. The highest number of search hits (public, community, household) can be placed low on the relationship ladder as they refer to groups of people. The other three labels (user, consumer and customer) can be placed on the higher end of the ladder as they refer to a more personal relationship (Figure 5.2). The size of the circles is proportionate to the number of mentions in journal abstracts. The larger the number of people who fit into the label, the lower the strength of the relationship. Referring to water users as the general public suggests a weaker relationship than referring to them as individual customers.

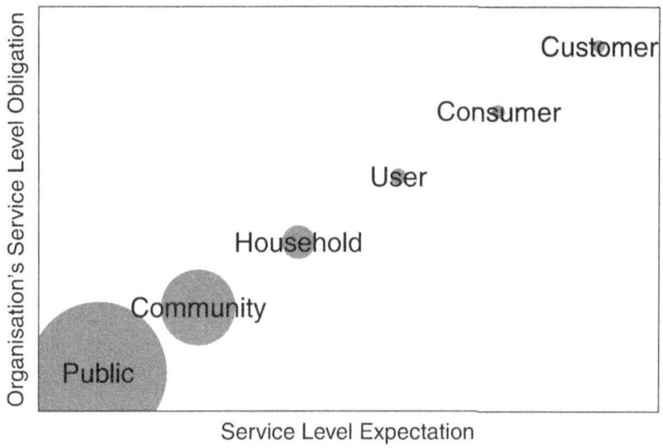

Figure 5.2 Relationship ladder of customer labels based on IWA journals.

This graph shows that collective labels are much more common in the IWA literature than individual customer labels. The collective labels emphasise the social benefits of water and the dominance of professional judgement while the singular labels emphasise the private benefits of water and consumer judgement (Laing, 2003). The more common use of collective labels over singular labels indicates a self-identification within the water utility industry as being a public service instead of a private service.

5.1.2.2 The beneficiary of a service

In S-D Logic, a service is defined as a process where one actor does something for another actor, the beneficiary (Lusch & Vargo, 2014). Within S-D Logic, the label customer is avoided by using the term beneficiary to indicate an actor who benefits from the service. This generic term is neutral as it does not provide any information about the relationship between two actors, like common customers labels do, but simply indicates the direction of the value proposition as indicated by the arrows in the value creation network. The definition of service in S-D Logic includes beneficiaries both internal and external to the organisation. This approach is common to internal marketing, where relationships within an organisation are characterised by defining an internal customer, as discussed in Chapter four (Lings & Greenley, 2010).

5.1.3 Customer segments

Segmentation is a foundational marketing strategy that organisations use to maximise the value they provide to customers. Marketing strategy can be placed along a spectrum from mass marketing to customised services for customers. Providing bespoke services to individual customers is only possible for luxury goods due to the high cost of customisation. Mass marketing is only suitable for commodities with few options for customisation. Water utilities tend to use a mass marketing approach as they provide the same undifferentiated service to all consumers, irrespective of their personal preferences. Although water utilities provide the same service to all their customers, segmentation enables organisations to target their supplementary services, such as payment, billing and providing information.

Segmentation is rarely discussed in the industry literature. Socio-economic factors and the related risk of financial hardship are common segmentation options for water utilities, also emphasised by the interviews discussed in Chapter three (CUAC, 2013; Hall Partners & Open Mind, 2011; Skellett, 1995). Another common segmentation method are the benefits that customers derive from the water such as irrigation, households, industry, sports grounds and so on.

Marketers use various methods to segment customers and selecting the best method depends on the organisation's objectives. Geographic segmentation classifies customers in accordance to their location, which makes sense for water utilities that manage geographically dispersed systems. This type of segmentation is relevant to water utilities due to the close relationship between the distribution network and geography.

Demographic segmentation groups customers based on factors such as age, gender and income, as is the case with specialised services for customers suffering financial hardship. Water utilities in countries where many people struggle to pay the bill use bottom-of-the-pyramid marketing strategies to provide services to the poorest customers. Focusing attention on the poorest segments has been beneficial to the people of the Philippines. Manilla Water found ways to connect more people to the network which brought costs down for millions of people. The company benefits economically from the increased volumes being sold and it creates public value by increasing the number of people with access to improved water (Coates *et al.*, 2004; Rangan *et al.*, 2011; Weidner *et al.*, 2010).

Lastly, psychographic segmentation classifies customers in accordance to their lifestyle and attitudes. For example, lifestyle segmentation in tap water could identify houses with gardens and those without into different segments. Research undertaken in the Australian state of Victoria has provided some insight into psychographic segmentation by investigating the attitude customers have towards water utilities (Burns *et al.*, 2011). This research was undertaken at the end the Millennium Drought and the segment descriptors therefore mainly relate to water conservation, as shown in Figure 5.3. The descriptors of these segments reveal a high level of disengagement of customers with water services, with the smallest segment being labelled as 'disempowered'.

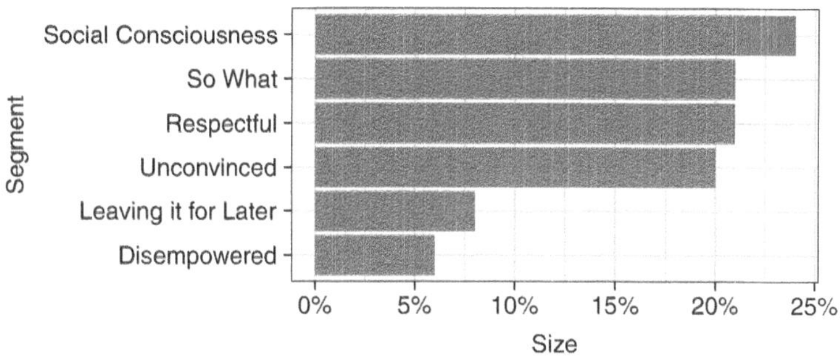

Figure 5.3 Residential water consumer segments. *Source*: Data obtained from Burns *et al.* (2013).

This segmentation was produced after Victoria just recovered from the Millennium Drought which is the reason that most lifestyle segments are related to saving water. The largest segment are people with a social consciousness who consider managing their water consumption to be a moral obligation. The segment labelled as 'respectful' are careful about their water consumption and value it as a resource but are less committed than the socially conscious consumers. An equally large segment are least likely to save water for the good of the community, labelled as 'so what'. One in five customers in Victoria are unconvinced about the importance of saving water and the activities that water utilities undertake. Some customers like to leave their concerns for later and are ambivalent about water services and are not well informed about what their water utility does. The smallest segment feels disempowered and is highly disengaged from water activities. This segmentation shows that more than half of tap water consumers in Victoria have a low level of involvement with their water services.

5.1.3.1 Implementing segmentation

The general principle of segmentation is that each customer segment has its own value proposition. Business customers will seek different benefits from their water service than urban customers do. In the example shown in Figure 5.3, the 'social consciousness' and 'respectful' water customers appreciate assistance from their water utility in saving water, while the other segments are less likely to seek this additional service.

There are no standard rules to determine which type of segmentation is the best for a water utility. Four criteria can be used to determine whether a segmentation model is likely to be beneficial. A segmentation strategy is only useful when each segment is measurable and has a substantial size. Most importantly, the segments have to be accessible. The utility needs to be able to identify which consumer is classified in which segment to provide a personalised service. This requires the service provider to maintain detailed information about each customer in their relationship management system. Segments also need to be differentiable in that they sufficiently differ from each other to warrant discrete services. Advanced statistical methods such as principal component analysis or clustering can be used to define segments that are significantly different from each other.

The Victorian segmentation model meets the first two criteria because the utility can use surveys to classify customers into each of the segments. These segments are also differentiable because they have been based on statistical analysis. However, the accessibility of these segments is problematic because it will be almost impossible to identify which customer in their service region belongs to which group unless all customers undertake a survey.

5.1.3.2 Persona's

The approaches to segmentation described in the previous section are based on statistics. A contemporary qualitative method of segmentation is to use personas. A persona is a description of a market segment as if it was a real person. Each persona has a detailed profile that lists their name, gender, their values, profession and hobbies and so on. The persona defines the archetypal customer by converting the otherwise lifeless data into a person. This method uses ethnographic research traditionally undertaken by anthropologists to define market segments. This research can take the form of interviewing people or observing them when they use the service (Revella, 2015). Personas should be based on research but can also emerge from the operational knowledge of employees that often interact with customers. Using operational knowledge is problematic for water utilities. Given the low frequency of contact between customers and their utilities, this knowledge might not be representative of the whole customer base.

The segments listed in Figure 5.3 can be converted to personas by describing them as real people with their own biography. For example, in the 'so what' segment we could find Darren, an Australian male in his forties with a graduate degree in economics. He and his family live in a suburban house with a garden. His job pays him a good wage and his water bills do not concern his family. He values a lush green lawn for his kids to play on and he is happy to pay for the water to keep it green. Such a persona profile can be extended by adding a photograph from a stock images provider. Personas go beyond the statistics by telling a story that helps the organisation to create service that maximises value in the lives of these people.

Using personas to segment customers is a powerful way to get to know your customers which can be used in internal marketing. The biographies can be used to assure that the people making decisions empathise with the customers impacted by their decisions. When, for example, a water utility discusses the option of introducing digital water metering, the project team can ask themselves what Darren or any other persona would think about the proposal. Personas can be used for role-playing sessions where employees step into the shoes of the customer by acting out their likely responses. This technique ensures that the decision process includes empathy for the points of views of the various customer segments.

5.1.4 Consumer involvement

Involvement is an important metric in marketing to describe the relevance a product or service has in somebody's life. People who own a car will most likely be highly involved with purchasing and owing the car due to the large amount of money involved and the social role it plays in developing their public self. Consumers will most likely have a much lower level of involvement with the instant coffee they drink than with the clothes they wear (Bienstock & Stafford, 2006; Flynn & Goldsmith, 1993). More formally, consumer involvement can be defined as 'a person's perceived relevance of the object based on inherent needs, values, and interests' (Zaichkowsky, 1985).

From a managerial point of view, involvement is important because it is causally related to willingness to pay and perceptions of quality. Consumers with a higher level of involvement are willing to pay more for a service and have a more favourable perception of quality (Cohen, 2000; Espejel *et al.*, 2009; Kinard & Capella, 2006). Understanding involvement in the context of urban water supply is also important because sustainably managing water as a common pool resource requires the active involvement of all users (Ostrom, 1990).

The level of consumer involvement depends on a complex array of factors, which are related to psychology, situational factors and the marketing mix of the service provider. The lowest level of involvement is considered a state of inertia which occurs when people habitually purchase a product without comparing alternatives. The highest possible level of involvement are the cult products where customers are fully devoted to a particular product or brand (Solomon *et al.*, 2010).

Commercial organisations use this knowledge to their advantage by maximising the level of consumer involvement through branding and advertising. This strategy is used effectively by the bottled water industry. Manufacturers focus on enhancing the emotional aspects of their product rather than on enhancing the cognitive aspects. Water utilities tend to use a reversed strategy and emphasise the cognitive aspects of tap water, the pipes, plants and pumps, rather than trying to create an emotional relationship with their consumers.

5.1.4.1 Involvement with tap water

The fact that tap water is an essential service where consumers don't have to shop around for alternatives suggests that consumers have a high level of service involvement. Contrary to this intuition, practitioner experience and the literature suggests that tap water is a service with a low level of involvement (Allon, 2006; Babakus, 1993; Fagan, 2011; Vloerbergh *et al.*, 2007; Watson *et al.*, 2002). In the interviews described in Chapter three, water utility customers were characterised as 'passive receivers of service', which suggests inertia towards tap water. The statement that tap water is a low involvement service seems paradoxical, given that access to tap water is a necessary condition for contemporary life.

Marketers can measure the level of involvement using psychometrically validated survey instruments. The most commonly used method is the Personal Involvement Inventory (Zaichkowsky, 1994). This is a two-dimensional scale consisting of cognitive involvement (importance, relevance, meaning, value and need) and affective involvement (involvement, fascination, appeal, excitement and interest). The survey instrument consists of ten semantic-differential items, shown in Figure 5.4. Customers are asked to position their views between two extremes such as Worthless and Valuable or Boring and Interesting. The level of involvement is scored between 10 and 70. This survey was completed by 797 water utility customers in Australia and the United States, which is described in detail in Chapter six.

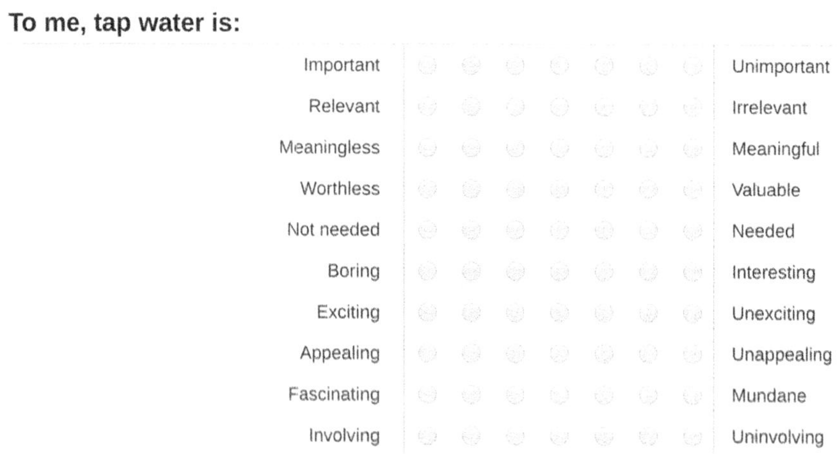

Figure 5.4 Involvement survey instrument. *Source*: Screendump of Qualtrics survey form.

Contrary to suggestions in the literature, the data shows that tap water is a service with a high level of involvement. The median level of involvement of 56 is located in the upper quartile of the involvement range (Figure 5.5). Comparing this finding with other research shows that the level of involvement for a car is higher than tap water, but the level of involvement for jeans, television and red wine is lower than for tap water.

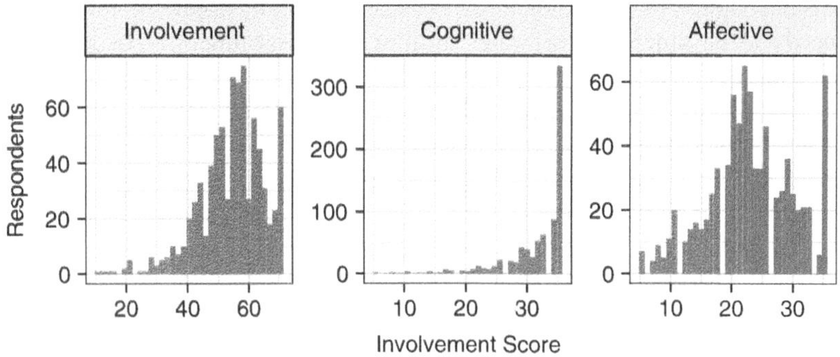

Figure 5.5 Personal involvement inventory for tap water.

Although respondents largely agree that tap water is important from a rational point of view, they are much less emotionally involved with the service. Their level of cognitive involvement ($Mdn = 23$) is significantly higher than their level of affective involvement ($Mdn = 34$, $p < .0001$). It is interesting to note that a large number of respondents reported the highest possible level of cognitive involvement.

Practical intuition leads to the paradoxical situation that consumer involvement for an essential product such as tap water is low. This paradox can be resolved by considering the large difference between the affective and cognitive levels of involvement. Tap water as an essential product will logically attract a very high level of cognitive involvement, because life in the developed world without it is unthinkable. However, as a unbranded, undifferentiated, monopolistic service, the level of affective involvement it attracts is significantly lower, as visualised in Figure 5.5.

5.2 MANAGING THE CUSTOMER EXPERIENCE

Customer experience management is a new direction in marketing that has emerged from practitioners that need to adapt to the realities of an evolving economy. Customers are no longer in the market for a product or a service but they seek experiences. The customer experience is the accumulation of a person's sensory, affective, cognitive, relational, and behavioural responses to a service. This experience is defined by a customer's 'journey' through touchpoints along the process of search, purchase, consumption. A touchpoint is a moment in the value creation process where the customer and the service provider interact. These touchpoints are continually judged against other experiences in a person's environment (Homburg *et al.*, 2017). The customer experience concept recognises that the relationship between a customer goes through phases and is characterised by all modalities of the human experience, both sensory and emotional.

This description of the customer experience generically applies to most services but needs to be amended for water utilities. The vast majority of touchpoints between the water utility and the customer is mediated by the customer's plumbing. Direct contact between the utility and its customers is rare as each consumer uses water several times per day but only contacts the utility when it is time to pay the bill, seek information or during supply disruptions. For water utilities to fully understand the experience, both the experience with core services and supplementary services need to be understood.

The fact that consumers use their own plumbing to mediate their experience with water demonstrates the concept of co-creation. Service failures related to issues with plumbing maintenance are a common occurrence (Cardew, 2000; Edwards, 2004; Espinosa-García *et al.*, 2014; Marchesan & Morran, 2004). Although utilities have little influence in how this plumbing is installed, customers attribute these failures in first instance to the service provider. Furthermore, the water quality paradox plays an important role in the customer's experience with tap water. The aesthetic properties of water provide tangible evidence of the service and are the primary determinant of the customer's experience. The public health properties of water are invisible to customers and cannot be experienced as such, as discussed in Chapter six.

5.2.1 Process mapping

The customer experience is not a static event but a process that starts before a person becomes a customer and only finishes after the customer is no longer associated with the service provider. One of the most common techniques to manage processes is to map these using the various techniques that are described in the literature. Processes mapping is a visual notation for depicting business processes using symbols that represent the actors and activities. Mapping techniques such as Business Process Modelling Notation (BPMN) and Unified Modelling Language (UML) are commonly used in business improvement initiatives. Process mapping is mainly used for designing improved supplementary services.

Service blueprinting is a technique that helps to manage the process of service delivery. The difference between service blueprinting and other process mapping techniques is that service blueprints are developed from the customer's perspective. A service blueprint is a process map that shows the actions of customers and those of the service provider. Activities are separated in the blueprint based on weather they are visible or not to the customer. The blueprint also indicates the physical evidence related to the service (Bitner *et al.*, 2008; Lovelock *et al.*, 2007).

Journey mapping is a recent model of the customer experience that has emerged from user-interface design in digital marketing. This method analyses the complete customer life cycle, as is the case in service blueprinting. Journey mapping also includes the thoughts, feelings and the emotional experience of the customer. Just like the service blueprint, a journey map includes all interactions and touchpoints between the customer and the utility. The affective component of the map identifies pain points in the customer's journey, which are moments where the customer is likely to experience difficulties or negative emotions. Journey maps can be developed in many different ways, with some creative aesthetically pleasing examples available on the internet. The best way to create a journey map is to involve all actors in the value creation process, including customers. Each perspective of the process needs to be incorporated so that any pain points can be identified and improved (Rosenbaum *et al.*, 2017).

The first step in improving services is to ensure that the current process is fully mapped, without yet including any future improvements. Only after the current situation is mapped should the team proceed to define improvements to alleviate pain points. An important factor in assessing the quality of a service process is the time cost to the customer. The lower the amount of time a customer needs to spend to pay a bill, lodge a complaint or any other service request, the higher the value the utility provides.

5.2.2 Managing complaints

Each interaction between a customer and their service provider is a moment of truth that can lead to either satisfaction or dissatisfaction. Every time a customer opens a tap, every time a customer interacts with a communication from the service provider, billing or otherwise, is a moment of truth. Most water utilities manage millions such moments of truth every day.

Due to the low level of personal interaction between water utilities and its customers, not much information is available about their perception of the service. Conducting market research is an expensive process which is only undertaken intermittently. This leads water utilities to rely on complaints from customers to obtain intelligence about their perception of service. An example of a service quality model that uses customer complaints is described in Section 6.4.1.

By relying on complaints, customers are effectively the last measurement instrument in the monitoring system and they act as sentinels to inform the utility of service failures. Unfortunately, complaints are often interpreted in a punitive way and performance indicators point towards minimising the number of complaints. This section argues that receiving a complaint is the best outcome for cases where a customer has a negative experience.

5.2.2.1 Complaint decision tree

Complaint research in the water utility sector is mostly limited to technical deliberations on which compounds in water cause a negative experience prompting customers to be unhappy enough to contact their water utility (Boxall & Prince, 2006; Dietrich *et al.*, 2014; Francisque *et al.*, 2011). More interesting than the physical aspects of customer complaints are the psychological and social aspects of complaint behaviour. Research in other service industries shows that complaint behaviour is a complex process that starts with the customer's expectations not being met. For example, customers expect water to be tasteless, but they notice an unpleasant taste when drinking from their tap. Perceiving a low level of service is, however, by itself insufficient for a consumer to lodge a complaint.

Customers are motivated to complain by a range of parameters, other than the physical quality of the water. When expectations are not met, customers make a series of choices, of which only one pathway leads to a complaint being lodged (Figure 5.6). The customer can decide to either take action, or not to take action. If action is taken, the customer chooses to take public or private action. Public action relates to lodging a complaint directly to the service provider or to external agencies, the media, or even taking legal action. Private action can take the form of changing service provider or spreading negative word of mouth (Day, 1977; Lovelock *et al.*, 2007). Moving to another water utility is not a viable option for customers, but they can choose to drink their own rainwater or bottled water instead of tap water, or even strive for self-sufficiency, leading to a loss in public image of the water utility. This decision tree shows that, besides avoiding water quality issues in the first place, a complaint by a customer is the most favourable option.

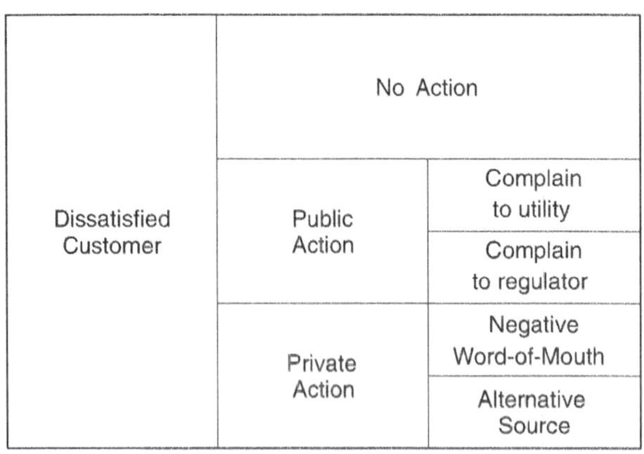

Figure 5.6 Post-dissatisfaction behaviour model.

5.2.2.2 *Likelihood of complaints*

The likelihood of each of these options occurring depends on a range of factors. The first influence on complaint behaviour relates to the nature and impact of the negative experience, determined largely by the chemical composition of the water. The level of dissatisfaction needs to be sufficiently high to motivate the customer to invest time in a complaint. The customer's threshold to lodge a complaint depends on individual circumstances that are influenced by the previous experiences a customer had with the utility, or for customers without direct experience with the service provider, by the public image of the utility.

The level of dissatisfaction is balanced with the cost of complaining. The financial cost of complaining is nil, but the time and psychological costs of complaining can be very high for customers. If complaints are not received with empathy as a default response, then the customer incurs a psychological cost. Speaking to a service operator who is unwilling to understand the customer's point of view is frustrating for customers.

The time a customer has to spend lodging a complaint relates to the access availability of communication channels. The cost of complaining also has to be matched by the resources the customer has to complain, such as computing and writing skills, ability to verbalise a complaint, access to the Internet or telephone and so on. The anticipated customer benefits of complaining relate largely to the image the customer has of the service provider. Customers who expect that complaining will not lead to a positive outcome are less likely to complain (Lovelock *et al.*, 2007).

These influencing factors are under some level of control by the utility. A service provider can lower the customer cost to complain by making multiple channels available through multiple levels of technology, and they can lower the psychological cost of complaining through an empathetic process. Some of these factors are, however, outside the control of the service provider and a background level of complaints in inevitable.

5.2.2.3 *Service recovery paradox*

The likelihood of lodging a complaint after a negative experience is quite low. In other industries it is estimated that only 5% of people who have a negative service experience always complain (TMI/SOCAP, 2005). A low level of complaints can counter-intuitively lead to problems because customers that don't complain will most likely resort to private action, which carries the risk of negatively impacting the public image of the service provider.

Marketing literature suggests that customers who experience a service failure but subsequently receive excellent service recovery, may ultimately be even more satisfied than they were before the failure. This has become known as the service recovery paradox. It should be noted that this phenomenon only occurs under best practice conditions. Customers will only be more satisfied after a complaint when the complaint process is managed perfectly. This situation only applies to small discrepancies between expectations and experienced service quality. For customers that experience a significant service failure, an excellent complaint process will be less likely to restore their trust in the organisation (de Matos *et al.*, 2007).

This phenomenon is as an opportunity to use the complaint handling process as a mechanism to convert customers, who would otherwise spread negative word of mouth, into advocates for the service provider. A best-practice complaint handling mechanism lowers the threshold for customers to complain, provides easy communication channels and provides an empathetic resolution experience. Introducing this approach will initially increase complaint levels, but will provide opportunities to leverage the service recovery paradox to improve the utility's image.

A minimum level of complaints in water utilities is unavoidable. The water quality paradox holds that customers focus on the aesthetic quality of the water, not on its ability to protect public health. The individual preferences of each customer cannot be met because the core service of water utilities is not differentiated. Once the utility decides to use chlorine or any other potentially controversial chemical there will always be a segment of the community that is dissatisfied with this decision. Research into complaints also shows that many of the reasons that customers complain cannot be controlled by the utility. A complaint is the best outcome for the utility because it provides intelligence on system performance. Complaints should thus not be treated punitively but should be received as a gift from the customer to the organisation.

5.2.3 Promoting the value proposition

Water utility customers have a low to medium level of affective involvement with tap water. Water utilities can maximise the level of involvement by implementing strategic marketing communications.

Water utilities often position themselves as technological organisations and show pride in their engineering achievements (Bell, 2012). The industry literature discussed in Chapter two confirmed that water utility professionals have a strong technological focus. The industry literature gives the impression that technological superiority is a sufficient condition for customer satisfaction. Communicating the technological benefits of water targets the cognitive level of involvement but ignores the affective component.

Water utilities provide an essential service to sustain life. A service is considered essential when there are no substitutes and when voluntary exit from the market is not possible (Ben-David, 2015). The review in Chapter one showed that neither of these conditions are fully met as several alternatives to tap water are available, albeit not competitively priced and at much lower levels of convenience.

The anthropological literature shows how water satisfies a far broader spectrum of wants and needs of consumers than mere survival. Expectations of *comfort*, *cleanliness* and *convenience* influence how consumers think about water (Shove,

2003). The value of water lies not in its essential nature of sustaining biological life, but in its role in shaping the social life of the community. Building on the first axiom of S-D Logic, water has no intrinsic value but acts as distribution mechanism to deliver the value proposition to consumers.

For managers, this understanding can influence the way water utilities position themselves within their community. Being aware of the intangible value of water changes how water utilities perceive and communicate their value proposition. The realisation that water has no intrinsic value, and that it is a vehicle to deliver value propositions which are only realised in interaction with customers, shifts attention from engineering to customer experience. In this way, engineering becomes a boundary condition for delivering value propositions instead of the prime focus.

The images in Figure 5.7 have been designed to illustrate how the value of tap water can be communicated beyond referring to technology by emphasising the intangible benefits of water. Having a shower is not just a way to maintain hygiene, but is also an inspirational experience that helps customers to be creative. Bathing your child is an emotional experience, the value of which lies far beyond the economic value of the water. The third image shows the value of water as the ubiquitous ingredient in food preparation which is not only needed for survival but a way to meet social needs. And lastly, the addition of fluoride provides public health benefits, which utilities shy away from promoting due to pockets of resistance within the community; this in contrast with other food products that promote additives to differentiate themselves in the market.

Figure 5.7 Communicating the value of tap water. *Source*: Stock photos Dreamstime.com

5.2.3.1 *Social marketing*

One of the defining differences between water utilities and commercial service providers is that they regularly use demarketing strategies to manage the supply of water. Strategies of demarketing are essentially the same as for regular marketing. Service providers can use pricing mechanisms, promotion and manipulating the supply chain to reduce demand for a product (Chipp & McKay, 2002; Cullwick, 1975; Kotler, 2011; Kotler & Levy, 1971; Lowe *et al.*, 2014).

Social marketing is a promotion strategy commonly used in countries with highly variable climates, such as Australia and the United States, to influence consumer behaviour and reduce demand for water (Lowe *et al.*, 2014; Madill & Ziegler, 2012; Walton & Hume, 2011). Although social marketing has been related to water consumption since its early inception (Kotler & Zaltman, 1971), theory on reducing household water use is sparse (Lowe *et al.*, 2014).

Social marketing exists downstream and upstream of the customer. Within the context of water utilities, downstream social marketing relates to activities to reduce water consumption. Fluoridation of drinking water to improve dental health is an example of upstream social marketing as societal objectives are not progressed through behavioural change, but through policy change independently of consumer choice (Hastings *et al.*, 2000).

Two types of strategies are available to influence water consumption: a structural and a voluntarist approach. Structural approaches change the conditions under which consumers are supplied to affect their behaviour. Using the elasticity of the price of water to moderate demand by increasing prices (Widrick, 1985), and using legally enforceable water restrictions are the most commonly used structural approaches. Using pricing and water restrictions to influence water consumption usually elicits strong social and political opposition due to the essential nature of the service.

Voluntarist approaches are at the core of social marketing and achieve behaviour changes by promoting ecological citizenship, aiming for a permanent attitude change (Lowe *et al.*, 2014). Water conscious consumption is influenced by perceptions of drought severity, the perceived importance of water conservation and the perceived impact of individual behaviour on achieving societal objectives (Das *et al.*, 1996). The effectiveness of voluntarist social marketing campaigns to reduce water consumption is an open question. Walton and Hume (2011) reported that consumers in the Australian state of Queensland not only changed their water conservation behaviour but also their accompanying attitudes. Conversely, Lowe *et al.* (2014) reported that the effect of household size, income, water price and environmental beliefs strongly affect water use while the effect of social marketing approaches appeared uncertain.

The results of the involvement research mentioned in Section 5.1.4.1 shows that campaigns to motivate customers to use less water should be aimed at increasing the affective involvement of customers. The cognitive involvement of customers is already at a very high level due to the essential nature of the service. The level of cognitive involvement is significantly lower because water is an unbranded and undifferentiated commodity. Marketing communication can be used to increase the affective involvement of customers.

5.3 THE INVISIBLE WATER UTILITY

A service is a process that is simultaneously consumed and produced, rendering the producer and the consumer inseparable. The co-creation concept recognises that the service provider creates value together with the consumer and that consumer behaviour influences the quality of the service (Cermak *et al.*, 1994; Grönroos, 1990; Heidenreich *et al.*, 2015; Roggeveen *et al.*, 2012).

Reticulated water core services are provided at the consumer's home through equipment owned by the customer, that is taps, toilets and the associated plumbing. Reticulated water services do not require any interaction between the customer and any of the utility's equipment or employees. In contrast to the core services, supplementary services require a high level of interaction between service provider and consumer, but these occasions are rare over the life of the customer relationship (Burns *et al.*, 2011; Skellett, 1995).

5.3.1 Time price of water

The co-creation of services implies that consumers sacrifice time to use water, in addition to the fee paid for the service. One of the causes for the growth of services in the developed world is the willingness of consumers to outsource their non-discretionary time in order to obtain additional discretionary time. Using time to enjoy a service is an opportunity cost to the consumer. Instead of using their time for a mundane activity, they could use it for something that is more valuable to them.

A service provider can increase value to the consumer value by minimising the amount of time required to enjoy the service (Becker, 1965; Davies & Omer, 1996). Nordhaus (1997) investigated the time-price of illumination through human history. From building open fires in prehistoric times to light bulbs in more recent times, the time-price for illumination, measured in hours work per Lumen hours, has reduced dramatically over the course of human history. The time price for electric illumination has been reduced to flicking the switch, paying a monthly bill and occasionally replacing the light. This concept can also be applied to reticulated water services (Williams & Mountjoy, 2012).

In contemporary reticulated water systems, the time-price incurred by consumers approaches is in principle equal to the amount of time required to use the water. The time price for water is approaching zero and only becomes noticeable in the case of service failures or inefficient facilitating services. When water is not available or not safe to drink, consumers incur a large time-price because they have to find water from alternative sources or boil contaminated water (MacGillivray, 2014;

McCoy, 2014). Imposing restrictions on water consumption in times of scarcity is another example of imposing a time-price as consumers need to find alternative ways to achieve the benefits they normally obtain from water use (Brennan *et al.*, 2007).

The principle of a time-price for water is dramatically illustrated in arid developing countries. In these countries women labour hours every day to obtain their family's daily amount of water. The enormous time-price of water for these societies places a great psychological and economic burden on these societies (Brown, 2010a; Stevenson *et al.*, 2012).

The time price illustrates the value created by utilities. They provide convenience by minimising the time-price to the only the time it takes to consume water plus the time required to use necessary supplementary services. Minimising the time-price by providing adequate pressure, high quality water and minimising the need to use supplementary services are central to the value proposition of water utilities.

The time price to consumer can be measured by the amount of times they need to contact their utility about their core service. Old research conducted by Wessex Water shows that customers contact their utility once every four to five years about a billing problem and once every ten to fifteen years about a service problem (Barrett, 1992). The lower the contact frequency for non-value-adding activities, the lower the time cost and the higher the level of value provided to customers.

5.3.2 Customer relationships and invisibility

If a perfectly operating water utility minimises contact with its customers, then it will be difficult to raise the level of affective involvement which impacts the utility's public image. The solution to this issue is that utilities not only provide services to their customers but also provide, as public service providers, value for the community as a whole.

Water utilities can enhance the customer experience an build strong relationships by developing a positive presence in the community through promotion, public relations and having a presence at public events. Many examples of public participation by water utilities have been published in the literature. Examples are free drinking water at events and public taste testing sessions. The invisibility requirement only applies to the core and facilitating services but not to the enhancing services.

Core services are the reason organisations are in business, i.e the supply of water. Facilitating services are closely related to the core service and enable their consumption, such as billing and complaint management. The enhancing services deliver additional value to the customer and they differentiate service providers from each other, such as support for public events. The relationship between core, facilitating and enhancing services is explained in more detail on page 129.

Chapter 6

Measuring the customer experience

The previous chapters have shown how value is not created by the water utility through purification and pressurisation of water but through a process of co-creation between the service provider and the customer. The value of water can only be expressed in the experience that customers have with their water. The experiences of customers are largely related to their interactions with the utility through their plumbing. Customers only occasionally contact their utility to manage administrative issues or to notify the service provider of any negative experiences. The total customer experience consists of the cumulative experiences with a water utility through their plumbing, as a member of the community and when using supplementary services. This chapter provides a method to measure the quality of these experiences using a psychometrically validated survey tool.

The experience of the customer is a complex phenomenon that starts with the objective physical qualities of the water, but is a subjective construct influenced by social and psychological forces. The phenomenology of water consumption is complex because objective physical parameters cannot be directly translated to the subjective customer experience. A consumer enjoying a bath might experience this as an event where she contemplates her life or generates creative ideas. Expressing this in physical parameters such as pH, turbidity and so on does not do justice to the customer's experience. Consumption of tap water is shaped by social expectations of cleanliness, pleasure-seeking behaviour and by the psychology of the consumer. The experience of tap water is thus defined by a complex a combination of factors, dominated by subjectivity.

doi: 10.2166/9781780408675_123

The customer's experience with a service can be measured using the concept of service quality, which is a measure of how the customer perceives the level of quality provided by the utility. The concept of quality is a prime consideration in the management of water utilities, as illustrated by the centrality of this topic in industry discourse, discussed in Chapter two. The water management literature is dominated by discussions on how to ensure customers have access to the right quantity and quality of water. These discussions are more than often written from the perspective of the water professional. The quality of services is measured in cubic metres, milligrams per litre and other physical parameters. The word quality has a very specific meaning in tap water services. Water quality is generally defined as a bundle of product attributes related to the level of risk to public health and the aesthetic appeal of water (National Health & Medical Research Council, 2011).

In the absence of competition and defecting customers, service quality is a defining performance criterion for water utilities. For publicly owned organisations, in absence of a profit motive, the quality of the service is a final measure of performance. Regulators specify performance indicators in their reporting requirements. As part of these regulations, water utilities are also often required to conduct customer surveys and report on the outcomes to regulators (Braadbaart, 2007).

Several performance measurement models for water services have been reported in the technical literature on water supply (Berg, 2010; Cabrera, 2011). These methodologies are focused on professional judgements instead of the views of customers. A study by Rambo *et al.* (2003) developed a model for customer service targets by assessing customer perspectives. Their work acknowledges that the value of water is not only located in its physical quality but also in its symbolic value.

The dominance of the physical sciences moves performance measurement to more objective technical measures as the subjectivity of customer observations is perceived to be undesirable. This preference is related to the difficulties that water utility professionals have with interpreting subjective information caused by a lack of social science competencies in water utilities (Meissner, 2015; Sofoulis, 2010).

The ISO 24510:2007 standard on performance measurement for water utilities provides guidance on both the technical and functional aspects of water utility performance. This standard presents a customer-focused view of service quality and recommends practitioners to consider aspects such as billing frequency and accuracy, both the health and aesthetic quality of water, and standards for complaints management (International Organizational for Standardization, 2007). This standard provides a range of focus areas for service quality but does not prescribe on how these measures should be developed.

Over the past three decades, marketing scholars have developed various models to measure service quality. Some of the most often cited models are the SERVQUAL and SERVPERF measurement scales, which are both based on psychometric measurement theory. Service quality is a construct of the mind that can only be measured by asking customers a series of questions that relate to their experience with the service or product. The ServQual and ServPerf scales have been mentioned in industry literature (Babakus, 1993; Vloerbergh *et al.*, 2007), but no evidence has been found of these being implemented in practice.

This chapter develops ServAqua which is a conceptual model to measure quality in water utilities based on the principles of Service-Dominant Logic (S-D Logic). The model consists of the hypothesised dimensions of technical and functional quality, as assessed from two perspectives: quality viewed intrinsically by the service provider and extrinsically by the customer. The first section defines the concept of quality overall and how it relates to S-D logic. This theoretical model is used to develop a service quality scale for water utilities. This chapter closes with examples on how to apply these theoretical concepts to water utility practice by combining physical measurements and customer observations into an indexed water system performance report.

6.1 DEFINING QUALITY

Quality can be easily understood from a common-sense perspective but it is a notoriously hard concept to define. Quality is an elusive construct for which a range of mutually exclusive definitions, each with its own variations, exists. Water utility professionals associate the word quality mostly with the objective physical quality of water. From a marketing perspective is a broader concept that also includes the subjective experiences of customers. These experiences can include using the tap water or interacting with employees. The quality of water consists of both the objective physical parameters of its pressure and purity, as well as the subjective psychological dimensions of the aesthetics of water.

The concept of quality is complex because of the wide variety of definitions. To illustrate the diversity of views on quality, Garvin (1988, p. 217) provided five types of definitions. These definitions are based on either 'innate excellence', 'quantity of some ingredient or attribute possessed by products', 'conformance to specifications', 'conformance at an acceptable price or cost' or 'consumer preferences'. This plurality of views does not stem from a lack of consensus among scholars but emphasises that quality can be studied from different perspectives. Some of these perspectives are metaphysics (quality as innate excellence), production management (quality as conformance to specifications, conformance at an acceptable price or cost) and marketing, which views quality as an expression of consumer preferences (Steenkamp, 1989).

Quality is most commonly defined with respect to the quality of *widgets* and their conformance to specifications, but can also be expressed as the experience of a *moment of truth*, the moment when the consumer compares the expected service with the actual service. Each of these perspectives is equally valid and the position one takes in the quality debate is determined by what the organisation seeks to achieve. Quality as innate excellence is suitable for marketing communications. Quality as conformance to specifications is important for product development and manufacturing. In a service delivery environment, the perspective of the customer is paramount.

The word quality is only ephemeral and illusive when used as an adjective, for example 'this is a high-quality hotel'. Used in this way, quality refers to innate excellence, without being specific about what that status entails. Statements using the word quality as an adjective are devoid of information as they cannot be empirically corroborated. This view of quality, while appealing to common sense, is unable to illuminate what the quality of a market offering entails and can only be used as a rhetorical device in marketing communication. The concept of quality becomes clearer when used as a noun, for example 'empathy is a desirable quality' or 'neutral taste is a quality of good water'. Using quality as a noun uncovers the dimensions of the quality of a market offering. The concept of quality can best be understood as a collection of qualities (empathy, responsiveness, taste, pressure and so on) that describe the market offering. Some of these qualities can be objectively measured, while others only reside in the mind of the observer.

This linguistic vignette shows that quality is not a phenomenon that can be perceived or measured independently, but a multivariate construct that can only be known indirectly by accounting for each of the qualities of the market offering. The quality of a product or service consists of one or more qualities that can either be measured (conformance to specifications) or subjectively reported by customers. From a marketing perspective, both dimensions are considered equally valid as they provide information about organisational objectives.

6.1.1 Quality, satisfaction and value

The customer experience is a holistic concept that is related to customer satisfaction, service quality and the perceived level of value. Quality is conceptually closely related to customer satisfaction and value. Customer satisfaction is the result of a customer's perception of value, where value is defined as perceived service quality relative to price. Both service quality and value are thus the cause of satisfaction. Favourable service quality and value perceptions directly influences customer satisfaction (Soutar, 2013).

Strongly cognitively oriented service quality and value appraisals lead to greater affective feelings of satisfaction (Cronin *et al.*, 2000). The perception of value held by customers is based on the perceived quality of the results of the service and how these have been delivered, balanced against the aggregated costs to the customer of availing themselves of the service (Walker *et al.*, 2006). Value is an expression of the benefits derived by the beneficiary, while quality relates to the what and how the service is performed. Value is thus the mediating factor between service quality and satisfaction, as visualised in Figure 6.1. These logical relationships show that service quality is a useful measure of value creation (Macdonald *et al.*, 2011).

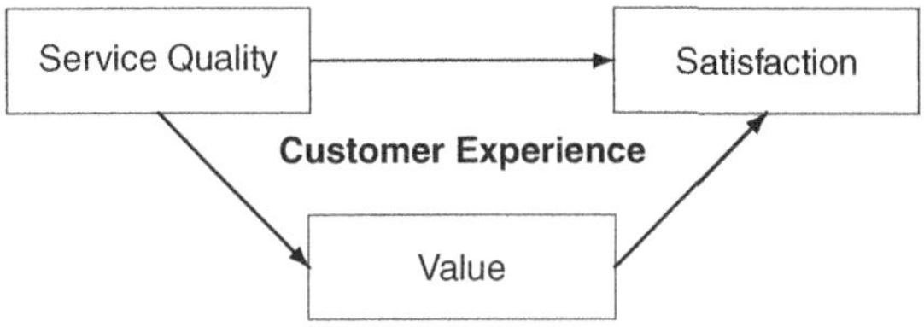

Figure 6.1 Relationship between quality, satisfaction and value.

6.1.2 Quality and service-dominant logic

Following the S-D Logic view of marketing, the physical quality of a product is only one perspective on the customer experience. As products are mechanisms to deliver a service, the subjective quality of the experience of the customer should also be taken into account. Early theories of quality revolved around the concept of product quality through physical measurements and conformance to specifications. Phrases such as 'doing it right the first time', 'conformance to requirements' or 'zero defects' used to be common in management literature (Garvin, 1988; Parasuraman *et al.*, 1985). The quality of services cannot be simply measured by comparing them with specifications. Due to the intangibility of services, quality is more difficult to evaluate than their physical aspects (Grönroos, 1990; Lusch & Vargo, 2014; Parasuraman *et al.*, 1985).

Service quality is a measure of the 'consumer's overall impression of the relative inferiority/superiority of the organisation and its services' (Bitner & Hubbert, 1994, p. 77). The service quality construct is an 'explicit model of how the quality of a service is perceived and evaluated by consumers' (Grönroos, 1984, p. 3). Service quality cannot be determined through objective physical measurements, as is the case in products, because compliance with specifications does not guarantee that individual customers evaluate the service favourably.

Referring back to the previous chapter, product quality is the domain of the engineer, while service quality is the domain of the marketer. Each of these views is based on a different paradigm of quality, with product quality aligned with the physical sciences and service quality with the social sciences.

To define quality within S-D Logic, the concepts of product quality and service quality need to be merged. Although the service aspects of water are dominant, this does not imply that product quality is irrelevant. Products function as the vehicle to provide a service to consumers and the quality of the physical product impacts the experience of the customer as much as the process of service delivery. Traditionally, contrasted concepts of product and service quality need to be merged into an overarching model for Service-Dominant Quality. The scope of S-D Quality needs to account for both products and services. It needs to incorporate objective physical quality as well as subjective service quality.

6.1.3 Perspectives of service-dominant quality

Service quality is defined from the consumer's perspective. However, viewing a service exclusively from the perspective of the customer provides only a partial understanding of quality. This is especially important in water services where most consumers do not have the expertise to judge the physical quality of the water but is able to judge its aesthetic quality.

The S-D Logic paradigm stresses the importance of co-creation between customers and service provider. Value is created through the interaction of the customer with the value proposition (Vargo & Lusch, 2004). Co-creation places limitations on the traditional views of customer centricity because it requires more than one party to create value. Balanced centricity extends the co-creation axiom by expressing the idea that all stakeholders in the value creation process have the right to satisfaction of needs and wants, as discussed in the previous chapter (Gummesson, 2008).

While in traditional quality frameworks, either the intrinsic view of the service provider or the extrinsic view of the customer is dominant, balanced centricity requires both perspectives to be given equal consideration. Relying on intrinsic quality alone prejudices the view of quality as the objective opinion of the scientist and the engineer. This technical orientation needs to be enhanced with the customer's extrinsic view of quality to measure service quality holistically. Both the perspective of the service provider and that of the customer are equally important. Recognising intrinsic quality does not negate the importance of the customer's point of view. Relying on extrinsic quality alone fails to accommodate the importance of attributes intrinsic to the service in influencing service quality and value offered to the customer (Burton *et al.*, 2001; Rust & Oliver, 1994; Walker & Johnson, 2009).

Service quality can thus be viewed from the perspective of both participants in the co-creation of value: quality as perceived and *extrinsically* attributed by customers and quality that is *intrinsic* to what is offered and how it is offered, as understood by the service provider.

6.1.3.1 Intrinsic service quality

Intrinsic quality is the perspective of quality from the service provider's point of view. Intrinsic quality is embedded in the delivery of the service or the product and intrinsically attributed by the service provider to the market offering. Intrinsic quality is based on management's perceptions of the expectations of consumers and their translation into specifications. The level of control over intrinsic quality is related to the level of tangibility of the service. The quality of physical widgets is easier to control than the quality of intangible moments of truth. In tap water, intrinsic quality involves all traditional measures of water quality such as laboratory testing and pressure monitoring. Intrinsic quality can be measured through standard quality monitoring processes and consists of a series of measurements of internal processes. (Collier & Bienstock, 2006; Coltman *et al.*, 2008; Walker & Johnson, 2009).

6.1.3.2 Extrinsic service quality

Extrinsic service quality is the level of quality as perceived by the consumer. On the consumer side, extrinsic service quality is defined by the gap between the expected service and the perceived level of service (Grönroos, 1984; Parasuraman *et al.*, 1985, 1988). Extrinsic service quality is the domain of traditional service quality theory and is commonly measured by surveying or interviewing consumers. Extrinsic service quality includes the invisible aspects of service provision and can be influenced by externalities, not under the control of the service provider. Consumers can also have a view of quality without having experienced the service, based on word-of-mouth and company image (Rust & Oliver, 1994).

Measuring extrinsic quality is complex because consumer attitudes are incorrigible. This means that they are only knowable to the person holding the attitude. An incorrigible statement is necessarily true and cannot be corrected by anyone. When a customer believes water to taste bad then this is an absolute statement that cannot be negated by providing alternative views. When customers complain about the taste or colour of water, utilities should not respond by stating that it is safe to drink but display empathy for the position that the customer takes (Goetz, 2014a). The incorrigibility of perceptions of quality is embedded in the fourth axiom of S-D Logic, which states that value can only be determined by the beneficiary (Lusch & Vargo, 2014; Vargo & Lusch, 2016).

The incorrigibility of attitudes towards quality raises barriers to our ability to know them. The most common method to assess attitudes is through psychometric theory. The basic premise of psychometric measurement is that the psychological state of a person expresses itself in the way a questionnaire is answered. The attitude itself is latent and cannot be measured directly, but we can measure the responses to questions caused by the attitude (DeVellis, 2011).

6.1.4 Scope of service-dominant quality

The experience that a customer has with the service offering extends beyond the actual product or service that is on offer. To analyse a service offering, it can be separated in core and supplementary services. Core services are the reason organisations are in business. Supplementary services consist of facilitating and enhancing elements. Facilitating services are closely related to the core service and enable their consumption. The enhancing services deliver additional value to the customer and they differentiate service providers from each other (Grönroos, 1990; Naipaul & Parsa, 2000). Following this distinction, the quality of a service can also be divided into the quality of the core service and the quality of the supplementary services. The core services require a different approach to measuring quality than supplementary services.

The Grönroos model for service quality distinguishes between the technical quality to describe the core service and functional quality to describe the supplementary services. Technical quality explains what service is provided through the measurable properties of the core service while functional quality describes the process of service provision. Technical quality is related to value because the core service is a conduit to deliver value to customers (Grönroos, 1984; Walker & Johnson, 2009).

6.1.5 Measuring service-dominant quality

The proposed model for quality under S-D Logic is visualised in Figure 6.2. Intrinsic service quality is measured using standard process control methods, while extrinsic quality relies on market research. A wide range of techniques is available to measure the technical quality of the core and supplementary services. Quality control techniques common to managing a water supply measure the technical quality of tap water. The supplementary services are measured through process control, such as a Grade of Service for a contact centre's responsiveness to service requests.

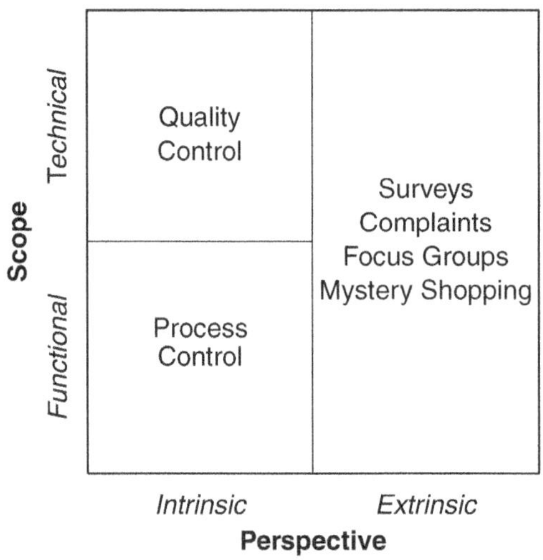

Figure 6.2 Service-dominant quality dimensions.

S-D Quality is always industry specific. The tangibility of the offering influences how technical quality can be measured. For example, the technical quality of a car is more related to engineering specifications while the technical quality of a hotel is much less based on physical measurements. Functional quality is more generic across industries because it relates to how the service is provided and often requires human interaction.

Although both perspectives on service quality are of equal importance, S-D Logic stipulates that value can only be determined by customers. This axiom places limitations on the way service providers measure intrinsic quality. A customer-centric water utility minimises any gaps between the intrinsic and extrinsic perspective. The service provider needs to develop measures for intrinsic quality that closely align with the views of the customer.

6.2 SERVAQUA: A SERVICE QUALITY INSTRUMENT FOR WATER UTILITIES

SERVAQUA is a survey instrument to measure service quality for water utilities, based on the theoretical model explained in the previous section. This model has been developed using data from customers in the United States and Australia (Prevos, 2013, 2016). This model has recently been confirmed using data from customers in the Mediterranean island of Malta (Psaila, 2017).

6.2.1 Core and supplementary services

The core services provided by water utilities are facilitated by a range of supplementary services. A generic model developed by Lovelock (1992) helps to better understand the various facilitating and enhancing services that water utilities offer. Facilitation services, necessary to deliver core services, are receiving orders, billing, payment facilities and providing information. Enhancing services which differentiate the service provider from competitors are consultation, hospitality, safekeeping and managing expectations.

Order taking is only applicable to reticulated water services at the time a new customer is connected to the system. Providing information is important as tap water is not packaged. Customers are informed about the pricing of water, water quality and other topics through a variety of communication channels, mainly through their regular bill. Billing of reticulated water services occurs at fixed intervals. Providing a water meter and recording the volume of consumed water are important aspects of this facilitating service. Payment of consumed water is the final facilitating service with special services for customers suffering financial hardship as an important aspect.

With respect to supplementary services, most contact between water utility customers occurs due to service failures. Hospitality and safekeeping only applies to customers entering the premises of the service providers, such as customer contact centres or other facilities. One aspect where safekeeping is of importance is the potential damage that water can cause to the customer's equipment, such as calcium deposits in appliances (Abeliotis *et al.*, 2015). Consultation is predominately related to providing technical advice to customers, for example information on how to reduce water consumption.

Table 6.1 summarises the various services that water utilities can provide to residential customers and to the community as a whole. It is based on service typologies published by the Water Research Foundation, industry research in Australia and the ISO 24510:2007 standard on water utility performance (Burns *et al.*, 2011; International Organizational for Standardization, 2007; Raucher *et al.*, 2012). This table only shows the direct customers and other beneficiaries, such as businesses. Other utilities and regulators have their own needs and wants and require special consideration. Research in Australia shows that the types of services that water utilities can provide are limited by the level of trust customers have in their utility (Burns *et al.*, 2011).

Table 6.1 Supplementary services provided by water utilities.

Service		Residential Customers	Community
Core	Water supply	In-home water filtration	Sale of bottled water
		Backflow prevention	Ice sales
	Sanitation		Medicine disposal
Facilitating	Ordering	New connections	
	Information	Consumption data	Educational resources
		Water saving information	Private source water testing
	Billing	Digital billing	
	Payment	Hardship assistance	
Enhancing	Consultation	Water audits	
	Safekeeping	Service line protection	
		Plumbing services	
		Leak detection	
	Hospitality	Contact centre	Event sponsorship
	Exceptions	Repairs	Emergency management

The higher the level of human interaction between the service provider and the customer, the more unpredictable the quality of service is and the harder it will be to manage this process. The core services of water utilities are very homogeneous because the physical quality can be controlled at a high level of reliability through technology. The supplementary services are much more subject to variability due to the higher level of interaction between employees and customers. Whereas the core service is undifferentiated with no possibility of customisation, supplementary services need to meet the individual requirements of the customer. When supplementary services are required, the otherwise anonymous customer that is served at arm's length, develops a direct relationship with the service provider.

6.2.2 Technical quality

The core service in reticulated water is dominated by tangible elements as water can be directly experienced by customers. The evaluation of the service is based on experience qualities because the customer is able to assess most aspects of the core service through visual, tactile, olfactory and gustatory verification. The customer can see, feel but hopefully not smell or taste the water. The very definition of water is that it is a tasteless and odourless liquid. Each opening of a tap consists of a moment of truth where the pressure, purity, and in some cases the taste of the water are assessed against expectations.

Customers cannot fully evaluate the technical quality of the service as some aspects of the purity of water are imperceptible. Customers can neither taste nor see pathogens and harmful chemicals. The imperceptible aspects of water quality are credence qualities and require the customer to trust the service provider. Although these qualities of the water cannot be directly verified by the customer's senses, they can be determined by the service provider. Extrinsic technical quality measures the perception consumers have on the quality of the water and intrinsic technical quality is measured through technological means.

6.2.2.1 The water quality continuum

The physical quality of water incorporates two points of view: *Good Water* as experienced by customers and *Safe Water* as determined by professionals (Parr, 2005). Safe Water relates to credence qualities as the safety of water cannot be directly discerned by consumers. Good water relates to experience qualities through the way the physical product is perceived by consumers.

Safe and good water are not diametrically opposed but are the extremes of a continuum. Safe water is expressed in the intrinsic objective quality of the water and good water in its extrinsic subjective quality. From the service provider's perspective, quality resides intrinsically in their decisions related to the delivery of what is offered to the customer. Safe water falls under the domain of engineering while good water falls under the domain of the social sciences.

The water quality continuum also illustrates the risk of tension between private and social benefits discussed in Chapter one. Safe water is a social benefit as judged by the professional while good water is a private benefit as judged by the consumer. This implies that safe water is mostly numerical and therefore predictable, while good water relies on parameters outside the physical quality of the water itself and is therefore unpredictable. The water quality continuum in Figure 6.3 helps visualise these differences.

Practical experience with water services reveals an inherent tension between intrinsic and extrinsic technical quality. Operators of drinking water systems are required to comply with local regulatory requirements. Meeting these requirements can, however, lead to a reduction in service quality. For example, adding chlorine is essential to ensure public health in that it destroys micro-organisms. In some communities, however, chlorine is perceived as an unwanted chemical, leading to a reduction in service quality (Kot *et al.*, 2011). The water quality paradox occurs when service providers neglect to minimise the difference between intrinsic and extrinsic service quality.

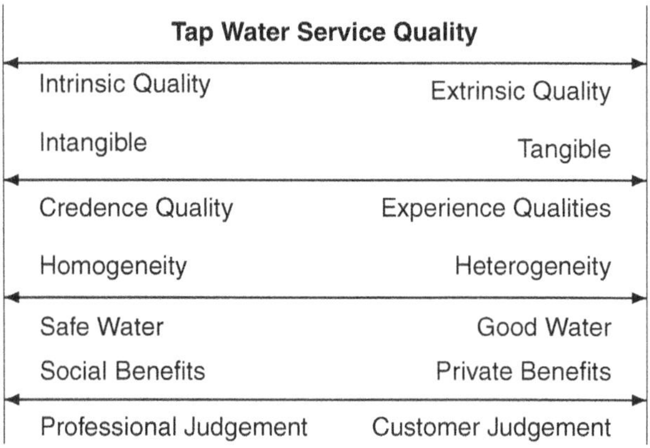

Figure 6.3 Water quality continuum.

The intrinsic technical quality of water is measured using incommensurable physical parameters, such as *E.coli* counts, salinity, *pH*, chlorine content and so on. Water quality indices have been developed to normalise water quality parameters into a single score to measure performance (Abbasi & Abbasi, 2012). Water quality parameters are mainly related to safe water and are able to report on the social benefits, but they fail to incorporate the private benefits of consumers by excluding the customer's experience of the service.

From the extrinsic perspective, technical quality can be subjectively assessed by consumers, just like the technical quality of a product. Within the tap water context, the technical quality of the service is determined by the experiential qualities of the supply, predominantly the absence of taste, odour or colour in the water, perceptions of continuity of supply, perceptions of safety and the right level of pressure. Extrinsic technical quality can be assessed through customer surveys or by analysing complaint statistics.

6.2.3 Functional quality

The extrinsic functional quality of water services is exclusively located in the supplementary services, which are characterised by a high level of customer interaction with the service provider and a low degree of tangibility. Babakus (1993) developed a scale to measure service quality for utilities which illustrates the importance of organisational behaviour in water utilities, discussed in Chapter four.

The model consists of fifteen elements related to billing, service characteristics, the behaviour of employees, and general questions:

- Billing clarity and accuracy
- Reliability and efficiency of services
- Customer service
- Safety consciousness
- Knowledge level of employees
- Providing services at the promised time
- Responsiveness
- Dependability
- Reassurance and understanding
- Having the interest of customers at heart
- Willingness of employees to assist customers
- Quality of management
- Politeness of employees

The ISO 24510:2007 standard for water utility performance management provides guidance to measure performance by water utilities based on the needs and expectations of water consumers. This document is the first part of a series of three. The remaining two provide specific guidelines for water and sanitation services. The areas of interest mentioned in this standard which covers supplementary services are in agreement with the scale developed by Babakus (1993), described above.

Many methods are available to measure intrinsic functional quality. Contact centres are most often guided by a Grade of Service, which is the percentage of contacts that are responded to within a given amount of time. Other internal performance indicators to measure the quality of service depend on the context of the service. A detailed discussion of these measures is outside the scope of this book.

6.2.4 SERVAQUA survey instrument

Water utilities regularly use customer surveys to find out what their customers think about the utility's brand, proposed pricing changes or investment proposals, satisfaction, and so on. People are regularly invited to complete surveys by either social researchers or commercial organisations so it is important that any survey is short and to the point. Many water utility surveys are developed by managers who often ask questions to satisfy their curiosity rather then developing a valid survey that informs decisions (Goetz, 2016).

Psychometrics is the science of measuring the attitudes of people using question banks and statistical methods to verify which questions are the most valid and reliable to ask of customers (DeVellis, 2011). An important principle in psychometrics and social science research is that an attitude, such as satisfaction or service quality, cannot be measured using a single question. The ubiquitous customer satisfaction question that asks customers to rate whether they are extremely dissatisfied, extremely satisfied or anything in between, or methods such as the Net Promoter Score, simplify the complexity of human psychology.

The ServAqua survey instrument has been developed based on the considerations described above. The instrument consists of a series of questions to measure service quality for water utilities. The questions for the technical and functional dimension builds on the water utility service quality scale developed by Babakus (1993), the interviews with customer advocacy groups, described in Chapter three, and the water industry literature content analysis outlined in Chapter two. These questions were administered to customers in Australia and the United States and the answers were psychometrically tested.

6.2.4.1 Technical quality survey questions

The technical quality dimension was measured using five questions. Interviewees mentioned the importance of supply reliability and the emphasis customers place on the experiential qualities of water supply (taste and visual appearance). The convenience aspect of tap water is associated with its availability, pressure and the perceived safety of the water. The items highlight the importance of time-price in water services through the words 'whenever' and 'always'. The technical quality items are formulated in absolute terms to ensure that the highest score entails a perfect level of service. Based on these considerations, the following five-item instrument was used:

t1 Tap water is available whenever I need it.
t2 My tap water is always safe to drink.
t3 My tap water is always visually appealing.
t4 My tap water always has a pleasant taste.
t5 My tap water always has sufficient pressure.

6.2.4.2 Functional quality questions

Interviewees often mentioned the quality of billing, both their accuracy and clarity, as an important determinant of quality. The knowledge level of employees and their empowerment to solve problems was also identified by interviewees as an important determinant of service quality. The responsiveness of service providers was seen as important in cases of service recovery. Another issue that was strongly represented in the interviews is the importance of water utility employees displaying empathy and their ability to understand the customer's needs.

Respondents repeatedly mentioned the fact that many water utilities are engineering focused organisations that occasionally lose sight of the customer perspective. One aspect mentioned several times by interviewees, but not forming part of the Babakus model, is the importance of providing timely and accurate information. The scale developed by Babakus (1993) largely corresponds with the data obtained in the exploratory phase. Two items related to the efficiency and level of safety consciousness were not used because these issues did not appear in the interviews:

- f1 My water bills are always accurate.
- f2 The services provided by my water utility are reliable.
- f3 My water utility always provides good customer service.
- f4 I can always depend on the services of my water utility.
- f5 Employees of my water utility have the knowledge to answer my questions.
- f6 My water utility consistently provides the services they promise.
- f7 Employees in my water utility give me prompt service.
- f8 When I have problems, my water utility is sympathetic and understanding.
- f9 My water utility has my best interests at heart.
- f10 Employees of my water utility are always willing to help me.
- f11 My water bills are easy to understand.
- f12 Employees of my water utility are consistently polite.
- f13 My water utility provides me with sufficient information.

6.3 VALIDATION OF THE SERVAQUA INSTRUMENT

The SERVAQUA instrument was validated using psychometric statistical analysis. This method tests the validity and reliability of surveys that aim to measure latent constructs such as personality, satisfaction, trust or in this case, service quality (DeVellis, 2011). Data was collected from four water utilities in Australia who invited their customers through email or social media to participate. Additional data was obtained by using an online survey panel in the United States, focusing on three major cities. The collected data is a convenience sample which did not contain demographic variables from respondents in order to keep the survey as short as possible and minimise uncompleted submissions.

Due to the broad geographical spread of potential respondents on the American survey panel, respondents were required to also complete two screening questions. The first question related to their place of residence and the second question asked whether they had tap water at home. Only customers located in Los Angeles, Denver or Boston continued to the next page of the questionnaire. The answer to the location question was verified using spatial metadata. The second page consisted of the consumer involvement instrument based on a scale developed by Zaichkowsky (1985, 1994). The involvement items closed with an open text item asking customers: 'If you have any additional comments about your views on tap water, please enter them below'. The involvement data is discussed in Chapter five.

The last page started with two items related to the customer's relationship with their service provider. Customers were asked to indicate whether they struggle to pay their water bills when they fall due, using a seven-point Likert scale from 'Strongly Disagree' to 'Strongly Agree'. The second question asked customers to indicate the frequency at which they contact their utility for support also using a seven-point Likert scale: 'Never', 'Less than Once a Month', 'Once a Month', '2–3 Times a Month', 'Once a Week', '2–3 Times a Week', and 'Daily'. The survey closed with the eighteen SERVAQUA items, which were measured using a seven-point Likert scale from 'Strongly Disagree' to 'Strongly Agree'. Technical and functional quality items were presented in the same matrix in randomised order. The final item of the questionnaire consisted of an open question which invited customers to provide additional comments about their tap water supplier.

6.3.1 Quantitative data

The Australian survey was submitted by 451 people between 2013 and 2015, of which 54 were excluded because they didn't provide consent and or did not answer any questions. The American survey was undertaken by 690 people in March 2015, of which 199 were terminated before completion. After cleaning the data, 797 respondents were used for analysis.

6.3.1.1 Customer characteristics

Both contact frequency and financial hardship were scored on the low end of the Likert scale (Table 6.2). The majority of customers do not experience financial hardship and rarely contact the utility. About 3.5% of respondents reported the highest level of difficulty paying their bills. There was no statistically significant difference between Australian and American respondents regarding financial difficulties or contact frequency.

Table 6.2 Financial hardship and contact frequency responses.

Aspect	n	Mean	SD	Skew	Kurtosis	se
hardship	797	2.64	1.81	0.84	−0.54	0.06
contact	797	1.63	1.03	2.74	9.50	0.04

6.3.1.2 Service quality

Responses to all technical quality items, except for t4 (taste), are highly negatively skewed ($skew < -1$), which indicates that the data is not normally distributed (Table 6.3). Some responses in the functional quality dimension (f2, f11) are also highly negatively skewed. The scores for the technical quality items were almost uniformly very high, which reflects the high level of reliability of water services in Australia and the United States. The item descriptions aimed to prevent extreme scores by using the absolute adjective 'always' but customer nevertheless mainly assigned high scores. It is interesting to note that functional quality scores are consistently lower than technical quality, which is a reflection of the lower level of process control over supplementary services due to the human dimension. The taste of water (t4) cannot be controlled to the same level due to personal preferences and thus shows a more normally distributed pattern.

Table 6.3 Descriptive statistics service quality items.

Item	n	Mean	SD	Skew	Kurtosis	se
t1	797	6.34	1.02	−2.47	8.12	0.04
t2	797	5.71	1.50	−1.40	1.40	0.05
t3	797	5.70	1.38	−1.40	1.84	0.05
t4	797	5.28	1.69	−0.98	0.06	0.06
t5	797	5.68	1.43	−1.40	1.56	0.05
f1	797	5.15	1.36	−0.49	−0.22	0.05
f2	797	5.80	1.19	−1.44	2.68	0.04
f3	797	5.14	1.33	−0.47	0.09	0.05
f4	797	5.65	1.24	−1.01	0.98	0.04
f5	797	5.06	1.29	−0.20	−0.30	0.05
f6	797	5.52	1.23	−0.77	0.36	0.04
f7	797	5.02	1.30	−0.17	−0.25	0.05
f8	797	4.78	1.36	−0.23	−0.02	0.05
f9	797	4.70	1.51	−0.39	−0.24	0.05
f10	797	4.99	1.33	−0.26	−0.20	0.05
f11	797	5.40	1.41	−1.00	0.67	0.05
f12	797	5.06	1.32	−0.33	−0.06	0.05
f13	797	5.17	1.40	−0.70	0.19	0.05

6.3.1.3 Psychometric analysis

Factor analysis was used to test the reliability and validity of the SERVAQUA scale. This analysis confirmed that the scale has a technical and functional dimension. In the final version, questions f2, f4, f6 and f11 were removed to improve the performance of the scale, leaving fourteen questions. Full details of the psychometric analysis can be found in the associated dissertation (Prevos, 2016).

To test the reliability of the survey instrument, the score for a sample of respondents was calculated by adding their responses to each question. Cronbach's-alpha was used to assess the reliability of the survey. An alpha value of 0.7 or higher indicates that the instrument explains more than half of the variance in the data which indicates internal consistency (DeVellis, 2011). Analysis showed that the individual factors and the total SERVAQUA instrument are highly reliable as they explain most of the variance in the data (Table 6.4).

6.3.1.4 Correlations between latent variables

Figure 6.4 visualises the distributions of technical quality, functional quality, customer characteristics and their correlations. All correlations between service quality dimensions and customer characteristics can be explained by theoretical considerations, which provides further support for the validity of the SERVAQUA measurement instrument.

Table 6.4 Descriptive statistics and reliability for each construct dimension.

	Items	n	Mean	SD	Skew	Kurtosis	se	Alpha
SERVAQUA	14	398	79.47	13.94	−0.32	0.18	0.70	0.92
Technical	9	398	34.55	6.19	−1.03	0.93	0.31	0.85
Functional	6	398	44.93	9.53	0.03	−0.26	0.48	0.93

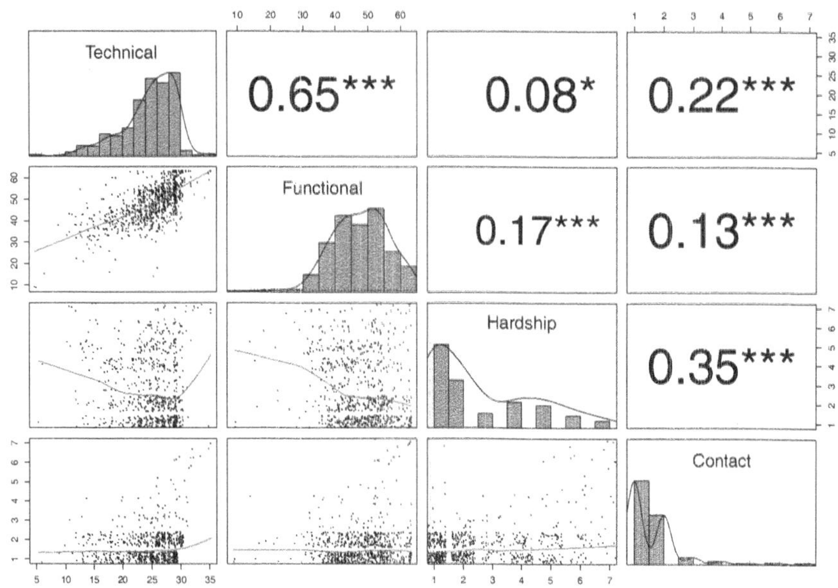

Figure 6.4 Correlations between service quality and customer characteristics.

Grönroos (1990) indicated that technical and functional quality are interrelated. Functional quality is more important to the perceived service than technical quality, assuming that the technical quality is provided at a satisfactory level. An acceptable level of technical quality is a prerequisite for a strong functional quality. The relationship is considered to be bi-directional because temporary problems with technical quality can be forgiven by the customer if the functional quality is at a high level. The two service quality dimensions correlate strongly with each other ($r(796) = 0.65$, $p < 0.001$), in line with these theoretical expectations (Grönroos, 1990).

The regular mention of the limited contact between the respondents and their water utility impacted their ability to score some of the functional quality items. Some respondents indicated to have selected 'Neither agree nor disagree' in the absence of experience with the utility's employees. This behaviour suggests that functional quality is related to contact frequency. Contact mainly occurs after service failures, which would suggest that contact frequency negatively correlates with both technical and functional quality. However, the data shows a strong positive relationship between contact frequency and technical quality ($r(796) = 0.22$, $p < 0.001$). The positive correlation between contact frequency and functional quality is also statistically significant ($r(796) = 0.13$, $p < 0.001$). These positive correlations can be explained by the service recovery paradox, which is a situation in which a customer's satisfaction after a service failure exceeds the level satisfaction before the failure occurred (de Matos *et al.*, 2007). Following the service recovery paradox, those water utility customers that contacted their utility after experiencing a failure are ostensibly more satisfied with the level of service quality than those who did not contact their service provider.

Financial hardship and contact frequency are also strongly positively correlated ($r(796) = 0.35$, $p < 0.001$). Customers suffering from financial hardship are more likely to require assistance from the service provider than those that have no problems paying the bills. This increased need for assistance does not translate into a higher level of perceived service quality. The level of financial hardship was found to be negatively associated with functional service quality ($r(796) = -0.17$, $p < 0.01$) and less so with technical service quality ($r(796) = -0.08$, $p < 0.05$). People who struggle paying their bills assessed the level of service as slightly lower as those who do not struggle with their financial obligations.

6.3.2 Qualitative data

The comments provided by customers support the findings from the interviews with customer advocates discussed in Chapter three. The taste of tap water was mentioned sixteen times, with all respondents providing negative feedback. One of the comments illustrates the tension between good water and safe water: 'the water supplier always claims the water is of excellent quality, but it does not taste that way'. The use of chlorine for disinfection was mentioned several times, mostly related to taste. One customer feared that the chlorine might adversely impact their health and was not satisfied with the scientific response from the water utility. Some customers also mentioned the addition of fluoride and expressed concern about its use, illustrating the water quality paradox: 'I do not believe in our water being fluoridated without my choice'.

Most positive responses were related to supplementary services: their promptness, politeness and responsiveness of employees. The comments highlight the critical nature of service recovery as many positive comments related to positive experience with the resolution of a problem they had, for example: '… it had been overflowing for a few days. I contacted [my water utility] and their response far exceeded my expectations. I subsequently hold [my water utility] in high regard'. There are also regular references to a negative experience with service recovery, mostly concerning the attitude of the utility employee, for example: 'spoke rudely, was impatient and talked over the top of me' and 'employees of my water utility weren't sympathetic when I had an underground leak'.

Several respondents pointed out that they found it difficult to answer the questions because of the limited contact: 'They supply the water, I pay the bill and that's all our relationship has been … therefore I cannot agree or disagree on some of the questions.' The lack of contact was not perceived as a negative and the invisibility principle was succinctly worded by one respondent: 'I've never had to contact them in ten years. That's how good the service is'.

The regular mention of the limited contact between utilities and their customers impacted the ability of customers to respond to some of the service quality items. Some respondents indicated to have used the middle score: 'Neither agree nor disagree'.

The qualitative information supports the existence of the water quality paradox and the tension between good and safe water. The tension between intrinsic and extrinsic quality, between the views of water quality held by customers and those held by the water utility are also expressed in this sample of customers.

6.3.3 Conclusions

The S-D Logic view of quality merges traditional views of product and service quality, which are usually perceived as diametrically opposed. Quality in S-D Logic consists of two dimensions. Scope is the part of the service offering being measured and perspective relates to either the customer's or the service provider's view of the service.

The scope of this quality model includes a technical and functional factor, each of which can be viewed from two perspectives: intrinsic and extrinsic. The intrinsic view is the perspective of the service provider while the extrinsic view is quality from the perspective of the customer. Following S-D Logic, the perspective of the customer has primacy, which implies that the way a service provider measures quality should as closely as possible replicate the customer's view of the world.

This conceptual model was tested with data obtained from water utilities customers in Australia and the United States by developing a measurement instrument for extrinsic quality of core and supplementary services. The instrument consists of five technical and eight functional quality items, which were tested for validity and reliability. These findings have also been confirmed with data obtained from water consumers in Malta (Psaila, 2017). The questions used in the ServAqua model can be used by water utilities to measure the level of service quality as perceived by their customers.

6.4 REPORTING INTRINSIC AND EXTRINSIC QUALITY

The survey method described in this chapter measures extrinsic service quality for both core and supplementary services, which is only one perspective of the SERVAQUA model. Measuring extrinsic quality is quite expensive and can only be undertaken intermittently using survey, focus groups or other social methods. Although the views of customers are the only views that matters with respect to measuring value, relying on this information alone is insufficient to manage a water supply. Managers need intrinsic information to ensure safe and good water is reliably delivered to customers.

Monitoring intrinsic service quality is essential intelligence to ensure customer value is maximised. Numerous methods exist to measure intrinsic service quality of water supplies, which fall in the domain of water quality expertise and not marketing. Intrinsic service quality can also include the voice of the customer. This can be achieved by translating the measure into the customer's perspective. For example, measuring the reliability of water supply is best expressed in customer minutes lost rather than the number of bursts per 100 km. Both parameters use the same data but reflect a different perspective on performance.

The question that arises from these reflections on performance measurement is how the intrinsic and extrinsic views of quality can be combined. How can the subjective view of customers be merged with the objective views of the water supply specialist? This final section briefly discusses a water system index based on the principles of SERVAQUA that combines intrinsic and extrinsic views on performance (Prevos, 2015).

6.4.1 Water system performance index

Communicating water system performance to an unqualified audience is a complex task. Whereas water usage is defined by a single volumetric parameter, water quality is a function of multiple incommensurable measures such as pathogen counts, *pH* and chemical concentrations. One popular method to achieve this is through a performance index which summarises the state of a water system within a single indicator (Abbasi & Abbasi, 2012).

Summarising performance of complex systems through an index is common practice in business management. Overall business performance is often communicated using a balanced scorecard, which integrates financial and non-financial measures in a cause-and-effect relationship (Kaplan & Norton, 1992). The performance of share markets is communicated using numerical indicators such as the Dow Jones index. Indices are also used to report performance in public policy, such as the Human Development Index or the Environmental Performance Index. The purpose of these indices is to combine related, but otherwise contrasting, information into a single ordinal number to facilitate an understanding of the subject.

Indices are also common in water management and a large variety of drinking water quality indices have been published (Abbasi & Abbasi, 2012). These indices focus on the biological and chemical properties of water but exclude the views of the users of water. Coliban Water has developed a method that creates an index for water system performance that combines intrinsic with extrinsic information (Prevos, 2015). This index visualises the performance of multiple systems within a service region using information from across the drinking water supply chain (Figure 6.5). The report combines information about Critical Control Points (CCP), laboratory testing, regulatory issues and customer feedback. The purpose of this report is to give the board of directors of Coliban Water visibility on the performance of water systems, including the views of customers. The combination of objective measurements and subjective customer complaints works because all data is reduced to a dimensionless index.

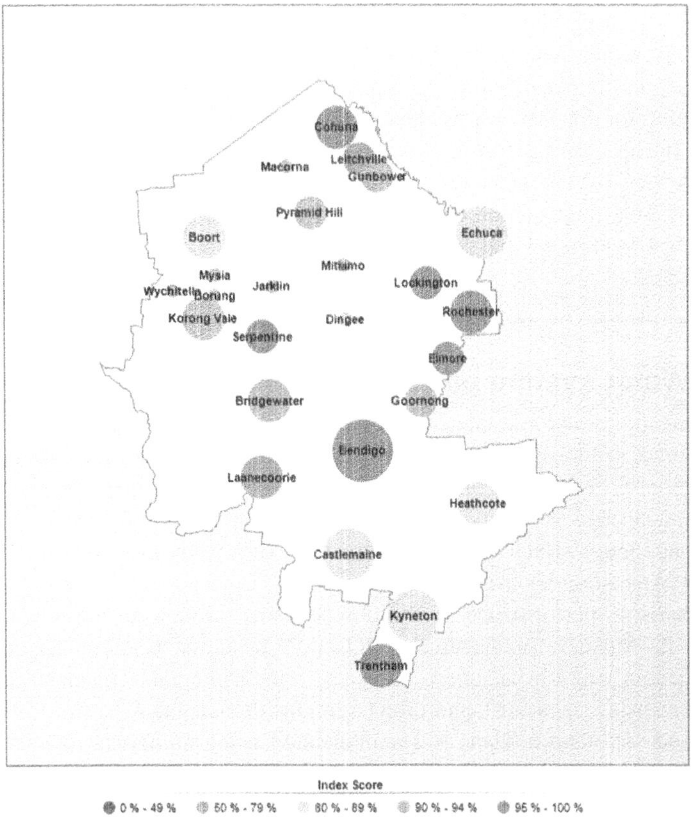

Figure 6.5 Coliban Water's Water system index (simulated data). *Source*: Prevos (2015).

To develop this system, forty water quality experts were asked about their views of the relative weightings of each of the factors in the performance index. Combining the objective information about physical performance with the subjective views of customers was seen as problematic by some of the respondents who commented that : '... the results will be ambiguous or inaccurate¡, 'not sure how this will help, seems a bit simplistic', and 'safety sometimes gets clouded by issues associated with customer aesthetic opinion.' These comments illustrate the professional orientation attitude discussed in Chapter four. Engineers and chemists are unaccustomed to analysing subjective data and prefer to rely on their scientific knowledge. The comment that public health is 'clouded' by the customer's perception of water quality reveals a strong level of technical orientation. Public health is a necessary condition for a high level of service quality but not a sufficient condition. Following balanced centricity, both the intrinsic and extrinsic view of quality are essential.

The water system index combines the perspective of the expert with the that of the customer in one index and traffic-light coloured map. This visualisation enables board members, who are not subject matter experts, to ask directed questions about the performance of water systems.

Chapter 7

Implementing water utility marketing

This book is based on a theoretical approach to marketing that explains how water utilities can be managed to maximise value to customers. Chapter one quoted psychologist Kurt Lewin who stated that nothing is more practical than a good theory. Theory does not have a high status with professionals as the term is often used in a negative way. This book is based on the idea that marketing theory provides a solid foundation for managers to make the best decisions.

Theories about business, either management, marketing or strategy, are almost always normative in nature. These are theories that describe how managers *should* behave to achieve their objectives. Many shelves in bookshops are dedicated to books that tell managers to follow some simple steps to improve their organisation. This approach is often problematic as it implies that the managerial objectives are already embedded within the theoretical framework. For example, the famous book *Good to Great* by Collins (2001) describes how companies can improve their performance by following a series of deceptively simple steps. Performance is defined by Collins as increased shareholder value. The problem with normative theories is that they presuppose a definition of success which might not be applicable to the specific context in which a manager operates. The objectives of a public service provider, such as a water utility, differ from the generic operations of commercial business and need a specific theoretical model.

In engineering, theory plays a different role than in management. The theoretical foundations of water supply management don't describe how an engineer should design a reticulation system or treatment facility. The theories of physics, chemistry and biology that form the foundation of water management describe what actually

doi: 10.2166/9781780408675_147

occurs. Engineers and scientists can apply that knowledge to shape reality in such a way that business objectives are met. Good management theories should also be based on first principles. Theories based on the foundations of sociology and psychology assist managers to understand reality and shape it in a way that maximises value to all beneficiaries of the utility.

This book is based on the understanding that foundational principles should also be used in water utility marketing by applying the principles of Service-Dominant Logic (S-D Logic). This model does not describe what should occur when people and organisations create value but it explains what actually happens in the value-creation process. Strictly speaking, S-D Logic is not based on axioms as it is founded on decades of research in sociology and economics.

Distilling managerial relevance from theoretical research is not always a simple proposition. Marketing often suffers from implementation anxiety (Jaworski, 2011; Reibstein *et al.*, 2009; Sofoulis, 2010; Strandvik *et al.*, 2014). After reading a research report managers can find themselves asking: 'What do I do differently tomorrow?'. Organisations with a strong focus on physical science and engineering struggle with making practical sense of the kinds of recommendations made by social researchers. The high level of abstraction of marketing research is often difficult to translate into actionable recommendations for managers (Jaworski, 2011; Sofoulis, 2005).

Jaworski (2011) classified four types of research implications to help make sense of how theory relates to practice. He recognised that some marketing knowledge is suitable for immediate *action* or requires further maturation, while other research inspires deeper *thinking* about managerial practice. The thinking part of theory is important because it changes the way managers understand their activities. Thinking differently about activities engenders new values and at a later stage will influence the actions managers take. This approach acknowledges differences in the timing and nature of the practical impact of research. The nature of the impact either stimulates application (action) or contemplation (thinking). While the relevance of action items is self-evident, contemplation is not often recognised as a managerial impact.

Marketing is essentially a mental model of how an organisation provides value to its customers. Providing managers with a relevant mental model assists them to move beyond viewing marketing as something undertaken by the marketing department but as a ubiquitous strategic concern for all organisational functions (Strandvik *et al.*, 2014). These impacts can be either related to the present or affect the future of an organisation, visualised in the managerial relevance matrix.

The purpose of this book is to motivate managers to think differently about their role in a water utility but also to provide advice on how to act differently to maximise customer value. The managerial relevance matrix in Figure 7.1 classifies the ideas presented in this book. The first cell in the matrix contains the water utility Marketing Mix and the SERVAQUA survey instrument as actionable findings that provide short-term impact. The invisible water utility concept as a key performance indicator is an action with a longer timing of impact as it requires further research. The remainder of the managerially relevant findings of this book are conceptual ideas that help managers to better understand their activities. Understanding and recognising the intangible value of water can improve the way water utilities engage with their customers by changing the nature of interactions. Value creation networks have the potential to reshape the way water utilities organise themselves to maximise value for all beneficiaries, including the natural environment. The value creation network concept is also placed in the future as it requires further development.

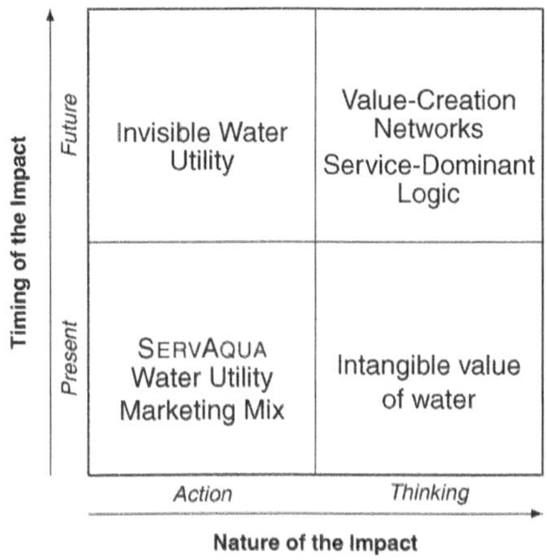

Figure 7.1 Managerial relevance matrix.

7.1 WATER UTILITY MARKETING SYNTHESIS

7.1.1 Marketing mix

The marketing mix presented in Chapter two is based on a content analysis of literature published by the International Water Association. This literature provides a comprehensive overview of the topics that are relevant to water utilities. The four dimensions derived from this analysis describe what type of activities water utilities undertake viewed through the lens of marketing theory. Although some topics, such as product quality, dominate the industry discourse, all four dimensions of the water utility marketing mix were given equal importance. This marketing mix model is not a rigid model and will most certainly not be the final word on what constitutes marketing of water utility services.

The idea of a marketing mix was first proposed in the literature by Culliton (1948) who coined the famous analogy between managers and a mixer of ingredients. Over the following decades, the marketing mix has become a rigid framework with many versions. Each author promotes their version as the most suitable model. Many scholars that write about the marketing mix forget that the original quote about mixing ingredients doesn't promote that managers follow rigid recipes:

> …the business executive is described in this book as a 'decider,' an 'artist' – a 'mixer of ingredients' who sometimes follows a recipe prepared by others, sometimes prepares his own recipe as he goes along, sometimes adapts a recipe to the ingredients immediately available, and sometimes experiments with or develops ingredients no one else has yet tried.

The marketing mix presented in this book should be viewed in the same way. While it is grounded in the reality of managing water utilities, it can by no means be viewed as a final truth. The purpose of defining the marketing mix for water utilities was to provide a mental model to shape the investigation into water utility marketing. What the content analysis of water utility journal article abstracts teaches is that marketing is not a specialised activity based on promotion and advertising. Marketing involves the whole of the water utility value chain, from catchment to the tap, and to the mind of the consumer.

7.1.2 Value proposition

Water utilities proud themselves on the engineering achievements that enable them to provide water to the communities they serve. Websites of many water utilities feature images of excavators and expensive equipment. This technology focus contrasts with websites from bottled water brands who promote their product with images that show either the pristine origin of the water or the intangible health and well-being benefits of drinking water.

The science of anthropology provides a unique insight into water consumption through participant observation. Chapter three summarises this literature which shows that water consumption has a strong social dimension. There is a large difference between the amount of water a human being needs to survive and the amount of water used by people with reliable reticulated water. Most water consumption is a discretionary hedonic consumption used to shape the social life of consumers. Enjoying a daily shower, for example, is not just a means to maintain hygiene but a ritual to create the public self.

Water utilities often place a value judgement on the different ways customers use water. Discretionary water use is labelled as a 'want' and only the water required to survive has the status of 'need'. From a marketing perspective, the service provider should not make judgements on the want or need for water. Humanist psychology suggests that all water consumption is based on intrinsic human needs. This can be either the need for self-actualisation by maintaining a well-manicured lawn, a need for social belonging when inviting friends to have a dip in the backyard pool, or a need for survival when using water to cook. This view of the value of water helps utilities better understand customers and shift their communication focus from engineering to people.

Understanding the human needs for water can help water utilities to develop a meaningful marketing mix. Many organisations interpret the value proposition as being a slogan that summarises the value the utility provides to customers. The concept used in this book is that the value proposition consists of three perspectives. Management designs the value proposition based on customer intelligence. Employees interpret this value proposition and implement it, and finally, customers convert the value proposition into value. The template described in Chapter three can help water utilities to design a management value proposition. This model focuses on the intangible value and is used as a template for internal marketing, to define service quality and manage the customer experience.

7.1.3 Internal marketing

In many jurisdictions, water utilities are being transformed from public authority to public service providers. An authority deals with ratepayers rather than with customers. This idea was supported by one of the interviewees discussed in Chapter three who stated that water utilities used to be managed by engineers, now they are controlled by economists and accountants and in the future they should be guided by anthropologists who understand how people behave.

Providing water services to communities relies on the skills and knowledge of highly specialised technical professionals trained in various fields of engineering and science. As discussed in Chapter three, this reliance on technology can lead to marketing myopia. This is a situation where a service provider has a false understanding of their value proposition. Other industries that rely on the specialised skills of engineers struggle with this issue and a lot of research has been conducted in the high-tech sector to better understand how engineering and science influences marketing objectives.

Chapter four summarises research undertaken in water utilities that investigates the attitudes of water utility employees. This research identified the concept of the professional orientation, which implies that employees are so occupied with their profession that they forget to understand the customer's perspective. This research also showed that mild tensions can exist between technical staff and their colleagues that deal with customers on a day-to-day basis.

Internal marketing uses the techniques of external marketing within the organisation to resolve any cultural issues that prevent the organisation from realising the value proposition. This can take many forms, such as employee rotation, training or using personas to bring the customer into the decision-making process. Chapter four discusses how one utility used taste testing by employees as a tool to engender empathy with the customer experience within the utility.

7.1.4 Service quality

Service quality is the subjective judgement that a customer has of the experience with their service provider. Water utilities excel at measuring their performance but do this mostly from their own internal scientific perspective. The current shift from asset performance to customer experience provides a new perspective on measuring performance. Existing performance indicators report on the intrinsic perspective of the utility, independent of its impact on customers. Knowledge about burst water mains is an important metric to optimise asset management, but if a pipe fails while everybody in the street is asleep, has it actually failed?

Chapter six discusses the SERVAQUA model for service quality that incorporates both the internal and external perspective of service quality. This model recognises the perspectives of both the customer and the utility have of quality. SERVAQUA also measures the complete suite of activities deployed by utilities. Technical quality measures the extent to which a customer's experience matches the expectation of a provided service. Functional quality measures how the service is provided and relates to the supporting services, such as billing and providing information.

The most practical way to measure service quality from the customer's point of view is using a survey. Chapter six discusses a survey instrument with 18 questions to measure extrinsic service quality following the SERVAQUA principles. This survey was administered to almost 800 people and their responses validated the theoretical assumptions on which this survey is based. The SERVAQUA can be used by water utilities to reliably measure extrinsic service quality.

The model SERVAQUA model is demonstrated with a water system performance index that incorporates both the view of the customer and the view of the service provider to ensure that the Board has visibility on system performance.

7.1.5 Customer relationships

Economies have been shifting from being product oriented to service economies. More recently it seems that economies are shifting from a service to an experience economy. Managing customer experiences is a relatively new field that combines knowledge from various other areas.

The definition of a customer for water utilities is not as straightforward as it is for commercial organisations. The value creation network for water utilities has many stakeholders, each of which benefits from the process in some shape or form. Chapter four discusses the idea that employees within an organisation can be viewed the same way as customers. The principle of balanced centricity suggests that all beneficiaries in the value creation network deserve to have their needs met. This concept goes beyond traditional customer centricity as it recognises that also regulators, employees and even the environment can be viewed as customers of water utilities.

Focusing on the traditional customer definition, water utilities practice a mass marketing approach where each customer is provided with the same type of service. Customer segmentation is a strategy that can help utilities to better understand their customers and maximise the value that is provided to each segment. Each significant segment of customers has a different value proposition as their needs and wants are different.

Water utility services are generally viewed as a low involvement service that is taken for granted by customers. Empirical research using respondents from two countries shows a very different pattern. The level of involvement in tap water, based on a standard survey instrument to measure this, is higher than most other products and services. This can be explained by the fact that water has an extremely high level of cognitive involvement due to the important role it plays in people's lives but a very low affective involvement as water is an unbranded commodity. Water utilities can increase the level of involvement by building stronger relationships with customers through communication and a high level of functional service quality.

7.2 WATER UTILITY MARKETING PARADOXES

Throughout this book, several areas of tension have been discussed that reveal the complexity and sometimes paradoxical issues faced by water managers. The paradox of value refers to the fact that the willingness to pay for water, which is essential to sustain life, is very low. The water quality paradox is a situation where the health and aesthetic properties of tap water are in conflict with each other. The involvement paradox arises because the level of involvement with water might seem to be very low but in actual fact is extremely high. Lastly, the invisibility paradox holds that the best level of service a water utility can provide is by being invisible to customers. This is an apparent contradiction because general marketing wisdom suggest that strong relationships with customers lead to better outcomes. These paradoxes illustrate the specific characteristics of water utility services and justify the idea that marketing theory needs to be amended to make it useful for this industry.

7.2.1 Paradox of value

The oldest paradox related to water was first formulated in the Bible and by Plato. The Bible book *Isaiah* (55:1) positions water as a human right, which aligns with the view that many people have in debates about water prices: 'Come, all you who are thirsty, come to the waters; and you who have no money, come, buy and eat! Come, buy wine and milk without money and without cost'. This invitation to share water with thirsty strangers illustrates the importance of water in arid cultures. Isaiah positions water as a right that is shared for free. Many people hold that water is a God-given right that should not have a price. The common counterargument to this position is that the water might fall free from the sky, but the pumps and pipes needed to deliver it to customers are not (Zetland, 2008). This paradox dissolves completely when water becomes sparse. In the words of Benjamin Franklin: 'When the well is dry, we know the worth of water'.

Some centuries later, Ancient Greek philosopher Plato wrote his *Euthydemus* (304B), in which his master Socrates says: 'For only what is rare is valuable; and water, which, as Pindar says, is the best of all things, is also the cheapest'. Pindar recognised that an essential substance such as water is available at a low price.

·The paradox of value also appears in the first book of modern economics. Adam Smith (2010) described the Diamond-Water Paradox in his book The *Wealth of Nations*: 'Nothing is more useful than water: but it will purchase scarce any thing; scarce any thing can be had in exchange for it. A diamond, on the contrary, has scarce any value in use; but a very great quantity of other goods may frequently be had in exchange for it'.

Although economists have resolved this paradox some time ago when they developed the theory of marginal utility, for water managers this paradox still plays very strongly as the willingness to pay for water is very low. The marketing solution to this issue is to emphasise the difference between value and price. The price we are willing to pay is related to the relative value we attach to a purchase. Water comes out of our tap, without effort and without any real involvement in the process. People are, however, highly involved with diamonds. Owning a diamond provides social status, access tap water is normalised and not a status symbol. Although water plays an important role in the social life of consumers, everybody in the community has access to the water, while not everybody has access to diamonds. The reason water demands a low price is because water is an essential service that is positioned by water utilities as a result of technology. Increasing the affective involvement customers have with tap water could play a role in increasing the willingness to pay.

7.2.2 Water quality paradox

Although in well-managed systems, the chances that somebody does not receive an excellent service through their tap are extremely small, consumer perception of water services is often not as good as engineers would hope for. There seems to be a paradox in the provision of water services.

Operators of drinking water systems are required to comply with regulations. Meeting these legal requirements can, however, lead to a reduction in service quality. For example, adding chlorine is essential to ensure public health in that it destroys micro-organisms. In some communities, however, chlorine is perceived as an unwanted chemical, leading to a reduction in service quality.

The perceptions of water quality held by consumers and water professionals are often very different. Consumers prefer good water and to focus on the tangible benefits through the aesthetics of water while professionals are more interested in the intangible public health benefits. Public health is not tangible to consumers because it is about the absence of illness. Indicators such as Disability Adjusted Life Years (DALY) are hard to explain even to water professionals.

The tension between the tangible and intangible aspects of tap water can lead to conflicts. Customers with limited trust in their service provider are likely to judge the safety of the water on taste, odour and colour, rather than relying on the information from the utility. This tension can lead to the water quality paradox, which is a situation when the health and aesthetic properties of tap water are in conflict with each other.

This tension is only a paradox if we consider that one view dominates the other. These sometimes contradictory views are both equally valid and cannot be resolved easily, just like light can be viewed as a wave and a particle at the same time, so are both safe water and good water equally valid views. Water utilities need to recognise these tensions and use marketing to either improve the status of tap water and recognise aesthetics as a necessary condition for customer satisfaction.

7.2.3 Involvement paradox

The general wisdom in water utility and business literature is that the level of consumer involvement in tap water is low. The fact that water is essential to life suggests that consumers of tap water have a high level of involvement with the service. This situation seems paradoxical because both assertions cannot be true at the same time. This paradox can be resolved by looking at involvement from two different angles. Tap water as an essential product will logically attract a very high level of cognitive involvement because life in the developed world is unthinkable without it. However, as a non-branded, undifferentiated, monopolistic service, the level of affective involvement it attracts will be significantly lower.

7.2.4 Invisibility paradox

Most of the direct experiences a customer has with their utility is through the facilitating services such as billing, payment facilities, information provision and service recovery. Given that these activities generally don't add value to the customer the number of interactions other then the core service should be avoided, which is the basic idea of the invisible water utility. This situation, combined with the low level of affective involvement, is problematic from a marketing perspective. Low involvement prevents the utility from developing a strong brand identity.

The Invisible Water Utility concept describes the highest possible level of Service Quality, that is minimising the amount of time people spend consuming water services. This implies that every single time a tap is turned on, expectations are met – no interruptions, no limitations of usage, no aesthetic issues, bills are accurate and easy to understand and so on. When achieving this ideal level of Service Quality, there is never a need for consumers to contact the utility, boil water to make it safe to drink, or to undertake other time consuming activities. The invisibility concept extends to the product quality of the water where the ideal service is one where the water is without colour or odour. The service provider thus becomes invisible to the consumer. Invisibility is in this sense a conceptualisation of consumer value.

The invisibility principle seems contradictory to the desire of water utilities to develop stronger relationships with their customers. The data obtained from customers supports the invisibility principle and that the quantity of engagement needs to be minimised. Further research into customer contact points in supplementary services can be conducted to more precisely specify the types of interactions and define engagement strategies. The engagement frequency needs to be minimised, but the quality of engagement should be maximised.

The relationship between customers and the water utility is, however, not the only relationship of importance in the value creation chain. The principle of balanced centricity is a useful departure from the customer focus rhetoric currently dominating industry discourse. The natural tensions between public and private benefits place limitations on the extent to which customer centricity can be implemented. Whereas commercial organisations aim to maximise customer centricity through market segmentation, positioning and targeting, the natural monopoly conditions of water utilities the impossibility of providing a differentiated water supply implies that total customer centricity is not achievable.

7.2.5 Marketing waste water services

This book has discussed the marketing of tap water services without giving consideration to its natural counterpart, the collection, transportation and disposal of waste water. The reason for this omission is that waste water services have different characteristics than tap water services. These differences requires sanitation to be approached differently from a marketing perspective. This final section discusses some of these differences and highlights any implications for the marketing sewerage services.

The value proposition of sewerage services is most likely to be very different from tap water. While tap water is a tangible product with service characteristics, sewerage is a pure intangible service. The value of sewerage services lies in the fact that they are totally intangible to the consumer. The service itself is performed by the sewerage system, imperceptible to the customer. Any perception of waste water service by customers is a service failure as the value of sanitation is defined by an increased distance between the waste and the community.

Sewerage services have a lot in common with the garbage disposal industry. The only difference between the two is that garbage is collected intermittently while sewerage is a continues service.

The intangibility of this service complicates developing a marketing mix for sewerage. In situations where both services are provided by the same firm, it would be advisable to interpret sewerage as a natural extension of tap water. In all instances where sewerage services are consumed, tap water is also consumed. Sewerage is as such positioned as an environmental service. Just like a manufacturer of computers can offer to recycle the machine, sewerage services can be positioned as an enhancing service for tap water. The water utility helps the customer to dispose of their waste in a manner that protects their health and that of the natural environment.

7.3 FURTHER RESEARCH

This book is not only defined by the theories, examples and topic described within it, but also by what has been omitted. For example, branding is one of the most ubiquitous topics in marketing. Although it is not discussed explicitly in this book, it is nevertheless as important to water utilities as it is for other types of organisations (Banerjee *et al.*, 2007; Dolničar *et al.*, 2014).

The managerial relevance of this book is generalisable to all reticulated water services as the structure of the industry is almost uniform in developed urban areas around the globe (Marketline, 2015). However, even though industry structure is uniform across the globe, perceptions of customers are influenced by their cultural background. Water consumption is largely impacted by cultural factors and cultural differences between service areas need to be taken into consideration. The results between American and Australian customers were not found to differ significantly, which could be explained by the fact that these cultures are highly comparable due to their shared cultural background (Hofstede, 2001). Future research could be conducted in non Anglo-Saxon countries to verify the generalisability of the findings.

In closing, the findings of this book can help water utilities to incorporate marketing knowledge into an industry otherwise dominated by the physical sciences. While good engineering is a boundary condition for providing value to customers, it is not a sufficient condition. The combination of engineering and marketing are not as incompatible as it seems. Marketing is after all, in the words of Kotler and Levy (1969): 'customer satisfaction engineering'.

References

Abbasi T. and Abbasi S. (2012). Water Quality Indices. Elsevier, Amsterdam.

Abeliotis K., Candan C., Amberg C., Ferri A., Osset M., Owens J. and Stamminger R. (2015). Impact of water hardness on consumers' perception of laundry washing result in five European countries. *International Journal of Consumer Studies*, **39**(1), 60–66.

Adams W. (2008). Major water Industry reform in Australia: a case study from Queensland. *Water Practice & Technology*, **3**(4), 1–11.

Addo-Yobo F. N. and Njiru C. (2006). Role of consumer behaviour studies in improving water supply delivery to the urban poor. *Water Policy*, **8**(2), 111–126.

Addo-Yobo F. N., Njiru C. and Sohail M. (2006). Determinants of households' intention to pay for improved water services: an application of the Theory of Reasoned Action. *Journal of Water Supply: Research and Technology—AQUA*, **55**(6), 419–425.

Ahmed S. (2009). Supply chain planning for water distribution in Central Asia. *Industrial Management & Data Systems*, **109**(1), 53–73.

Akpinar M. G. and Gul M. (2014). An assessment of consumer preferences on the drinking water market: today to the future. *Journal of Water Supply: Research and Technology—AQUA*, **63**(7), 525–531.

Allan R., Jeffrey P., Clarke M. and Pollard S. (2013). The impact of regulation, ownership and business culture on managing corporate risk within the water industry. *Water Policy*, **15**(3), 458.

Allon F. (2006). Dams, plants, pipes and flows: from big water to everyday water. *Reconstruction: Studies in Contemporary Culture*, **6**(3), 1–24.

Allon F. and Sofoulis Z. (2006). Everyday water: cultures in transition. *Australian Geographer*, **37**(1), 45–55.

American Marketing Association (2013). About AMA: Definition of Marketing.

doi: 10.2166/9781780408675_159

Ameyaw E. E., Memon F. A. and Bicik J. (2013). Improving equity in intermittent water supply systems. *Journal of Water Supply: Research and Technology—AQUA*, **62**(8), 552–562.

Anderson J. C., Narus J. A. and van Rossum W. (2006). Customer value propositions in business markets. *Harvard Business Review*, **84**(3), 90–99.

Angelakis A. N., Mays L. W., Koutsoyiannis D. and Mamassis N. (2012). Evolution of Water Supply Throughout the Millennia. IWA, London.

Appadurai A. (ed.) (1986). The Social Life of Things: Commodities in Cultural Perspective. Cambridge University Press, Cambridge.

Arnould E. J. (2007). Service-Dominant Logic and consumer culture theory: natural allies in an emerging paradigm. In: Consumer Culture Theory, number 11 in Research in Consumer Behavior, R. W. Belk (ed.), Elsevier, pp. 57–76.

Askew L. E. and McGuirk P. M. (2004). Watering the suburbs: distinction, conformity and the suburban garden. *Australian Geographer*, **35**(1), 17–37.

Auriol E. and Picard P. M. (2009). Infrastructure and public utilities privatization in developing countries. *The World Bank Economic Review*, **23**(1), 77–100.

Awofeso N. (2012). Ethics of artificial water fluoridation in Australia. *Public Health Ethics*, **5**(2), 161–172.

Babakus E. (1993). Measuring service quality in the public utilities. *Journal of Nonprofit & Public Sector Marketing*, **1**(1), 33–49.

Bagozzi R. (1974). Marketing as an organized behavioral system of exchange. *Journal of Marketing*, **38**(4), 77–81.

Bakker K. (2007). The "commons" versus the "commodity": alter-globalization, anti-privatization and the human right to water in the global south. *Antipode*, **39**(3), 430–455.

Banerjee A., McFarland D. A., Singh R. and Quick R. (2007). Cost and financial sustainability of a household-based water treatment and storage intervention in Zambia. *Journal of Water and Health*, **5**(3), 385–394.

Barber M. and Jackson S. (2011). Aboriginal water values and resource development pressures in the Pilbara region of north-west Australia. *Australian Aboriginal Studies*, **2**, 32.

Barnard H. E. and Henderson C. R. (1917). Public utilities and public health. *Journal American Water Works Association*, **4**(2), 218–222.

Barrett C. (1992). Putting customer quality first. *Managing Service Quality*, **2**(4), 215–219.

Barrett G. and Wallace M. (2009). Characteristics of Australian urban residential water users: implications for water demand management and whole of the system water accounting framework. *Water Policy*, **11**(4), 413–426.

Bastian H. (1998). Speaking up for ourselves: the evolution of consumer advocacy in health care. *International Journal of Technology Assessment in Health Care*, **14**(01), 3.

Becker G. S. (1965). A theory of the allocation of time. *The Economic Journal*, **75**(299), 493–517.

Bell S. (2012). Urban water systems in transition. *Emergence: Complexity and Organization*, **14**(1), 45–58.

Ben-David R. (2015). In Whose Long Term Interest is Customer Consultation by Water Authorities? Institute of Water Administration, Melbourne.

Berg S. V. (2010). Water Utility Benchmarking: Measurement, Methodologies and Performance Incentives. IWA Publishing, London.

Berkholz P., Stamminger R., Wnuk G., Owens J. and Bernarde S. (2010). Manual dishwashing habits: an empirical analysis of UK consumers. *International Journal of Consumer Studies*, **34**(2), 235–242.

Berkholz P., Kobersky V. and Stamminger R. (2013). Comparative analysis of global consumer behaviour in the context of different manual dishwashing methods: analysis of global manual dishwashing methods. *International Journal of Consumer Studies*, **37**(1), 46–58.

Bhatti M. and Church A. (2004). Home, the culture of nature and meanings of gardens in late modernity. *Housing Studies*, **19**(1), 37–51.

Bienstock C. and Stafford M. (2006). Measuring involvement with the service: a further investigation of scale validity and dimensionality. *The Journal of Marketing Theory and Practice*, **14**(3), 209–221.

Bijker W. E. (2012). Do we live in water cultures? A methodological commentary. *Social Studies of Science*, **42**(4), 624–627.

Bijker W. E., Hughes T. P. and Pinch T. J. (eds) (2012). The Social Construction of Technological Systems: New Directions in the Sociology and History of Technology. MIT Press, Cambridge.

Bitner M. and Hubbert A. (1994). Encounter satisfaction versus overall satisfaction versus quality. In: Service Quality: New Directions in Theory and Practice, R. T. Rust and R. L. Oliver (eds), Sage Publications, Thousand Oaks, London & New Delhi, pp. 72–94.

Bitner M., Ostrom A. and Morgan F. (2008). Service blueprinting: a practical technique for service innovation. *California Management Review*, **50**(3), 66–94.

Blazquez-Resino J. J., Molina A. and Esteban-Talaya A. (2015). Service-Dominant Logic in tourism: the way to loyalty. *Current Issues in Tourism*, **18**(8), 706–724.

Booms B. and Bitner M. (1981). Marketing strategies and organization structures for service firms. In: Marketing of Services, J. Donnely and W. George (eds), American Marketing Association, Chicago, IL, pp. 47 –51.

Borden N. (1964). The concept of the marketing mix. *Journal of Advertising Research*, **4**(2), 2–7.

Boxall J., Skipworth P. and Saul A. (2003). Aggressive flushing for discolouration event mitigation in water distribution networks. *Water Supply*, **3**(1–2), 179–186.

Boxall J. B. and Prince R. A. (2006). Modelling discolouration in a Melbourne (Australia) potable water distribution system. *Journal of Water Supply: Research and Technology—AQUA*, **55**(3), 207–219.

Braadbaart O. (2007). Collaborative benchmarking, transparency and performance: evidence from the Netherlands water supply industry. *Benchmarking: An International Journal*, **14**(6), 677–692.

Brennan D., Tapsuwan S. and Ingram G. (2007). The welfare costs of urban outdoor water restrictions. *The Australian Journal of Agricultural and Resource Economics*, **51**(3), 243–261.

Breslin P. A. (2013). An evolutionary perspective on food and human taste. *Current Biology*, **23**(9), R409–R418.

Brown R. (2010a). Unequal burden: water privatisation and women's human rights in Tanzania. *Gender & Development*, **18**(1), 59–67.

Brown R. E. (2010b). Business Essentials for Utility Engineers. CRC Press, Boca Raton, FL.

Bruvold W. H. (1970). Mineral taste and the potability of domestic water. *Water Research*, **4**(5), 331–340.

Burlingame G. A. and Mackey E. (2007). Philadelphia obtains useful information from its customers about taste and odour quality. *Water Science & Technology*, **55**(5), 257.

Burn S., Marlow D. and Tran D. (2010). Modelling asset lifetimes and their role in asset management. *Journal of Water Supply: Research and Technology—AQUA*, **59**(6–7), 362.

Burns A. V., Vardy S., Rowe M. and Murchinson R. (2011). Victorian Water Customer: Water Services Needs & Values Summary Report, Technical Report, Accenture, Melbourne.

Burton J., Easingwood C. and Murphy J. (2001). Using qualitative research to refine service quality models. *Qualitative Market Research*, **4**(4), 217–223.

Butler P. and Collins N. (1995). Marketing public sector services: concepts and characteristics. *Journal of Marketing Management*, **11**(1–3), 83–96.

Byatt I. (2013). The regulation of water services in the UK. *Utilities Policy*, **24**, 3–10.

Cabrera E. (2011). Benchmarking Water Services: Guiding Water Utilities to Excellence. IWA Publishing, London.

Camps L. M. (2000). Quality of service and customer satisfaction. *Water Supply*, **18**(1), 497–499.

Cardew P. T. (2000). Simulation of lead compliance data. *Water Research*, **34**(8), 2241–2252.

Carper K. L. (1991). Engineering Code of Ethics: beneficial restraint on consequential morality. *Journal of Professional Issues in Engineering Education and Practice*, **117**(3), 250–257.

Cermak D. S. P., File K. M. and Prince R. A. (1994). Participation in service specification and delivery. *Journal of Applied Business Research*, **10**(2), 90–97.

Chipp K. and McKay T. M. (2002). A Baseline for Demarketing Water in the Developing World: A Case Study of Gauteng, South Africa. Institute of Marketing Managers, Rosebank, Johannesburg.

Chisari O., Estache A. and Romero C. (1999). Winners and losers from the privatization and regulation of utilities: lessons from a general equilibrium model in Argentina. *The World Bank Economic Review*, **13**(2), 357–378.

Coates S., Sansom K., Kayaga S., Chary S., Narender A. and Njiru C. (2004). Serving All Urban Consumers: A Marketing Approach to Water Services in Low and Middle-Income Countries. Book3: PREPP Utility Consultation with the Uraban Poor. Water, Engineering, and Development Centre, Loughborough.

Cohen M. (2000). Consumer involvement–driving up the cost. *Consumer Policy Review*, **10**(4), 122–125.

Collier J. E. and Bienstock C. C. (2006). Measuring service quality in e-retailing. *Journal of Service Research*, **8**(3), 260–275.

Collins J. (2001). Good to Great: Why Some Companies Make the Leap and Others Don't. Random House, London.

Coltman T., Devinney T. M., Midgley D. F. and Venaik S. (2008). Formative versus reflective measurement models: two applications of formative measurement. *Journal of Business Research*, **61**(12), 1250–1262.

Constantinides E. (2002). The four S web-marketing mix model. *Electronic Commerce Research and Applications*, **1**, 57–76.
Cooper B., Burton M. and Crase L. (2011). Urban water restrictions: attitudes and avoidance. *Water Resources Research*, **47**(12), W12527.
Cornes R. and Sandler T. (1986). The Theory of Externalities, Public Goods, and Club Goods. Cambridge, Cambridge University Press.
Croke G. and Williams J. (2011). Customer consultation in water project planning and delivery: lessons from case studies. *Water Practice & Technology*, **5**(4), 1–8.
Cronin J. J., Brady M. K. and Hult G. T. M. (2000). Assessing the effects of quality, value, and customer satisfaction on consumer behavioral intentions in service environments. *Journal of Retailing*, **76**(2), 193–218.
CUAC (2013). Meaningful & Genuine Engagement Perspectives from Consumer Advocates, Technical Report, Consumer Utilities Advocacy Centre.
Culliton J. W. (1948).The management of marketing costs. Division of Research. Graduate School of Business Administration, Harvard University.
Cullwick D. (1975). Positioning demarketing strategy. *Journal of Marketing*, **39**(2), 51–57.
Cunha R. C. and Cooper C. L. (2002). Does privatization affect corporate culture and employee wellbeing? *Journal of Managerial Psychology*, **17**(1/2), 21–50.
Daniel K. (1998). *Determinants of Service Behaviour Among Contact Service Personnel*. PhD thesis, University of Technology, Sydney.
Darby M. R. and Karni E. (1973). Free competition and the optimal amount of fraud. *Journal of Law and Economics*, **16**(1), 67–88.
Das M., Das H. and Mackenzie F. (1996). Satisfaction with town services: do user experience and satisfaction count? *The Service Industries Journal*, **16**(2), 191–205.
Davies G. and Omer O. (1996). Time allocation and marketing. *Time & Society*, **5**(2), 253–268.
Day R. (1977). Extending the Concept of Consumer Satisfaction. *Advances in Consumer Research*, **4**, 149–154.
Deichmann U. and Lall S. (2007). Citizen feedback and delivery of urban services. *World Development*, **35**(4), 649–662.
de Matos C. A., Henrique J. L. and Alberto Vargas Rossi C. (2007). Service recovery paradox: a meta-analysis. *Journal of Service Research*, **10**(1), 60–77.
DeVellis R. F. (2011). Scale Development: Theory and Applications. Applied Social Research Methods, 3rd edn. Sage Publications.
Dietrich A. M., Burlingame G. A. and Hoehn R. C. (2003). What the nose knows. Strategies for taste-and-odor testing methods. *Opflow*, (October), 10–14.
Dietrich A. M., Phetxumphou K. and Gallagher D. L. (2014). Systematic tracking, visualizing, and interpreting of consumer feedback for drinking water quality. *Water Research*, **66**, 63–74.
DiLorenzo T. J. (1996). The myth of natural monopoly. *The Review of Austrian Economics*, **9**(2), 43–58.
Dolničar S. and Hurlimann A. (2009). Drinking water from alternative water sources: differences in beliefs, social norms and factors of perceived behavioural control across eight Australian locations. *Water Science & Technology*, **60**(6), 1433.
Dolničar S. and Hurlimann A. (2010). Water alternatives-who and what influences public acceptance? *Journal of Public Affairs*, **11**(1), 49–59.

Dolničar S. and Saunders C. (2006). Recycled water for consumer markets—A marketing research review and agenda. *Desalination*, **187**(1–3), 203–214.

Dolničar S. and Schäfer A. I. (2009). Desalinated versus recycled water: public perceptions and profiles of the accepters. *Journal of Environmental Management*, **90**(2), 888– 900.

Dolničar S., Hurlimann A. and Grün B. (2014). Branding water. *Water Research*, **57**, 325–338.

Doria M. D. F. (2006). Bottled water versus tap water: understanding consumers-preferences. *Journal of Water and Health*, **4**(2), 271–276.

Doria M. D. F. (2010). Factors influencing public perception of drinking water quality. *Water Policy*, **12**(1), 1.

Douglas M. and Isherwood B. C. (1996). The World of Goods: Towards an Anthropology of Consumption: With a New Introduction, rev. ed edition, Routledge, London, New York.

Doyle C. (2011). A Dictionary of Marketing. Oxford Paperback Reference. Oxford University Press, Oxford and New York.

Drucker P. (1954). The Practice of Management. Perennial Library, New York.

DuBrin A. J. (2005). Self-perceived technical orientation and attitudes toward being flattered. *Psychological Reports*, **96**(3), 852–854.

du Fossé I. (2010). The Marketer-Engineer Conflict in High-Technology Companies. Master of Supply Chain Management, Open Universiteit Nederland, Heerlen.

Dupont D. P. and Jahan N. (2012). Defensive spending on tap water substitutes: the value of reducing perceived health risks. *Journal of Water and Health*, **10**(1), 56.

Duruz J. (1994). Suburban Gardens: Cultural Notes. In: Beasts of Suburbia: Reinterpreting Cultures in Australian Suburbs, S. Ferber, C. Healy and C. McAuliffe (eds), Melbourne University Press, Melbourne.

Edwards M. (1995). Empowering the engineer to contribute to the marketing process. In: IEE Half-Day Colloquium: Engineering Marketing—Engineer and Marketeer in Perfect Harmony? IEE, vol. 1995, pp. 1–3.

Edwards M. (2004). Controlling corrosion in drinking water distribution systems: a grand challenge for the 21st century. *Water Science & Technology*, **49**(2), 1–8.

Elliott B. (2010). 'There's formaldehyde in my water': absurd discourse about the Sydney desalination project—A case study. *Asia Pacific Public Relations Journal*, **11**, 75–82.

Erlenkotter D., Sethi S. and Okada N. (1989). Planning for surprise: water resources development under demand and supply uncertainty: I. the general model. *Management Science*, **35**(2), 149–163.

Espejel J., Fandos C. and Flavián C. (2009). The influence of consumer involvement on quality signals perception: an empirical investigation in the food sector. *British Food Journal*, **111**(11), 1212–1236.

Espinosa-García A. C., Díaz-Ávalos R. and Solano-Ortiz R. (2014). Removal of bacteria, protozoa and viruses through a multiple-barrier household water disinfection system. *Journal of Water and Health*, **12**(1), 94–104.

Essential Services Commission (2012). Review of Water Performance Report Indicators—Staff Discussion Paper, Technical Report C/12/7501.

Estache A. and Rossi M. A. (2002). How different is the efficiency of public and private water companies in Asia? *The World Bank Economic Review*, **26**(1), 139–148.

Estache A., Perelman S. and Trujillo-Castellano L. (2005). Infrastructure Performance and Reform in Developing and Transition Economies: Evidence From a Survey of Productivity Measures, Technical Report, World Bank.

Euzen A. (2003). How do consumers perceive water quality? Example of an anthropological study carried out in Paris. *Water Supply*, **3**(3), 263–269.

Fagan B. M. (2011). Elixir: A Human History of Water. Bloomsbury, London.

Falahee M. and MacRae A. W. (1995). Consumer appraisal of drinking water: multidimensional scaling analysis. *Food Quality and Preference*, **6**, 327–332.

Falp S. and Le Masurier P. (2009). Understanding Customers' Views. Quantitative Research Into Customers' Priorities – Overall Report, Technical Report PR09, OFWAT.

Ferenstein G. (2015). Uber CEO Spells Out His Endgame, In 2 Quotes.

Festa G., Cuomo M. T., Metallo G. and Festa A. (2016). The (r)evolution of wine marketing mix: from the 4ps to the 4es. *Journal of Business Research*, **69**(5), 1550–1555.

Fife-Schaw C., Barnett J., Chenoweth J., Morrison G. M. and Lundéhn C. (2008). Consumer trust and confidence: some recent ideas in the literature. *Water Science & Technology: Water Supply*, **8**(1), 43.

FitzPatrick M., Davey J., Muller L. and Davey H. (2013). Value-creating assets in tourism management: applying marketing's Service-Dominant Logic in the hotel industry. *Tourism Management*, **36**, 86–98.

Flynn L. R. and Goldsmith R. E. (1993). Application of the Personal Involvement Inventory in marketing. *Psychology and Marketing*, **10**(4), 357–366.

Franceys R. (2006). Customer committees, economic regulation and the Water Framework Directive. *Water Science & Technology: Water Supply*, **6**(5), 9.

Franceys R. W. and Gerlach E. (2011). Consumer involvement in water services regulation. *Utilities Policy*, **19**(2), 61–70.

Francisco J. P. S. (2014). Why households buy bottled water: a survey of household perceptions in the Philippines. *International Journal of Consumer Studies*, **38**(1), 98–103.

Francisque A., Rodriguez M. J., Sadiq R., Miranda L. F. and Proulx F. (2011). Reconciling 'actual' risk with 'perceived' risk for distributed water quality: a QFD-based approach. *Journal of Water Supply: Research and Technology—AQUA*, **60**(6), 321.

Freeman M. C., Quick R. E., Abbott D. P., Ogutu P. and Rheingans R. (2009). Increasing equity of access to point-of-use water treatment products through social marketing and entrepreneurship: a case study in western Kenya. *Journal of Water and Health*, **7**(3), 527.

Frow P. and Payne A. (2011). A stakeholder perspective of the value proposition concept. *European Journal of Marketing*, **45**(1/2), 223–240.

Fumagalli E., Garrone P. and Grilli L. (2007). Service quality in the electricity industry: the role of privatization and managerial behavior. *Energy Policy*, **35**(12), 6212–6224.

Fuss N. and Stamminger R. (2012). Application of best practice tips in manual dishwashing in Germany and Spain: best practice tips in manual dishwashing. *International Journal of Consumer Studies*, **36**(2), 173–182.

Gallagher D. L. and Dietrich A. M. (2014). Statistical approaches for analyzing customer complaint data to assess aesthetic episodes in drinking water. *Journal of Water Supply: Research and Technology—AQUA*, **63**(5), 358.

Garrick D., Bark R., Connor J. and Banerjee O. (2012). Environmental water governance in federal rivers: opportunities and limits for subsidiarity in Australia's Murray–Darling River. *Water Policy*, **14**(6), 915.

Garvin D. A. (1988). Managing Quality: The Strategic and Competitive Edge. Free Press, New York.

General Assembly (2010). The Human Right to Water and Sanitation, Technical Report 64/292, United Nations, New York.

Gentina E., Palan K. M. and Fosse-Gomez M.-H. (2012). The practice of using makeup: a consumption ritual of adolescent girls: adolescent girls' ritual use of makeup. *Journal of Consumer Behaviour*, **11**(2), 115–123.

Gleick P. (1998). Water in crisis: paths to sustainable use. *Ecological Applications*, **8**(3), 571–579.

Glennon R. (2005). Water scarcity, marketing, and privatization. *Texas Law Review*, **83**(7), 1873–1902.

Goetz M. K. (2014a). Communicating Water's Value: Talking Points, Tips, and Strategies. American Water Works Association, Denver, CO.

Goetz M. K. (2014b). Perception and the value of water: it's all relative. *Journal AWWA*, **106**(6), 62–65.

Goetz M. K. (2016). Why surveys work and how they can disappoint. *Journal AWWA*, **108**(5), 70–74.

Goetz M. K. (2017). Communicating Stormwater, Wasterwater, & Watersheds Water's Value. Communicating the Value Part 2. American Water Works Association, Denver, CO.

Goetz M. K. and Prevos P. (2016). Tap Crawl from Down Under. *Journal—American Water Works Association*, **108**, 47–53.

Gorre-Dale E. (1992). The Dublin Statement on Water and Sustainable Development. *Environmental Conservation*, **19**(02), 181.

Goubert J.-P. (1989). The Conquest of Water: The Advent of Health in the Industrial Age. Polity Press, Cambridge.

Grafton R. Q. and Ward M. B. (2008). Prices versus rationing: Marshallian surplus and mandatory water restrictions. *Economic Record*, **84**, S57–S65.

Grasso D. and Burkins M. B. (eds) (2010). Holistic Engineering Education: Beyond Technology. Springer, New York.

Gribble S. J., Scott D., Mawdesley M. and Al-Jibouri S. (2006). Learning to be real engineers, architectural engineering and design management. *Architectural Engineering and Design Management*, **2**(1–2), 101–114.

Griffin A. and Hauser J. R. (1992). Patterns of communication among marketing, engineering and manufacturing—A comparison between two new product teams. *Management Science*, **38**(3), 360–373.

Grönroos C. (1984). A service quality model and its marketing implications. *European Journal of Marketing*, **18**(4), 36–44.

Grönroos C. (1990). Service Management and Marketing: Managing the Moments of Truth in Service Competition. Lexington Books, Lexington, Mass.

Grove S. J., Fisk R. P. and John J. (2000). Services as Theater: guidelines and implications. In: Handbook of Services Marketing and Management, T. A. Schwartz and D. Iacobucci (eds), Sage Publications, Thousand Oaks, pp. 21–36.

Gummesson E. (1995). Relationship marketing: its role in the service economy. In: Understanding Services Management: Integrating Marketing, Organisational Behaviour, Operations, and Human Resource Management, W. J. Glynn and J. G. Barnes (eds), Wiley, Chichester and New York, pp. 244–268.

Gummesson E. (2008). Extending the service-dominant logic: from customer centricity to balanced centricity. *Journal of the Academy of Marketing Science*, **36**(1), 15–17.

Gupta A. K. and Wilemon D. (1991). Improving R&D/marketing relations in technology-based companies: marketing's perspective. *Journal of Marketing Management*, **7**(1), 25–45.

Gupta A. K., Raj S. P. and Wilemon D. L. (1985). R & D and marketing dialogue in high-tech firms. *Industrial Marketing Management*, **14**(4), 289–300.

Gussy M., Gold L., Riggs E. and Waters E. (2008). Community water fluoridation: is it still worthwhile? *Just Policy*, **47**, 14–21.

Habel N., Maddox M. and O'Donoghue M. (1993). Myth, Ritual and the Sacred. University of South Australia, Underdale.

Hall D., Lobina E. and de la Motte R. (2004). Making Water Privatisation Illegal: New Laws in Netherlands and Uruguay, Technical Report, Public Services International Research Unit.

Hall Partners and Open Mind (2011). Customers of Water and Energy Providers in Financial Hardship: A Consumer Perspective, Technical Report 3981, Essential Services Commission, Melbourne.

Hanemann W. M. (2005). The economic conception of water. In: Water Crisis: Myth or Reality? P. P. Rogers, M. R. Llamas and L. Martinez-Cortina (eds), Taylor & Francis, London and New York, pp. 61–91.

Hardin G. (1968). The tragedy of the commons. *Science*, **162**(3859), 1243–1248.

Harrison J. and Stamp W. (1991). The sydney water board: a cause study in the development of a marketing and customer service focus. In: Service Delivery and Public Sector Marketing, C. O'Faircheallaigh, J. Wanna and P. M. Weller (eds), Macmillan Co. of Australia, South Melbourne.

Hart N. and Stapleton J. (2012). The CIM Marketing Dictionary. Taylor and Francis, Hoboken.

Hart O., Shleifer A. and Vishny R. (1997). The proper scope of government: theory and an application to prisons. *Quarterly Journal of Economics*, **112**(4), 1127–1161.

Hastings G., MacFadyen L. and Anderson S. (2000). Whose behavior is it anyway? The broader potential of social marketing. *Social Marketing Quarterly*, **6**(2), 46–58.

Heidenreich S., Wittkowski K., Handrich M. and Falk T. (2015). The dark side of customer co-creation: exploring the consequences of failed co-created services. *Journal of the Academy of Marketing Science*, **43**(3), 279–296.

Hirschman E. C. and Holbrook M. B. (1982). Hedonic consumption: emerging concepts, methods and propositions. *Journal of Marketing*, **46**(3), 92–101.

Hoekstra A. Y. (1998). Appreciation of water: four perspectives. *Water Policy*, **1**, 605–622.

Hoekstra A. Y. (2013). The Water Footprint of Modern Consumer Society. Earthscan, from Routledge, London, New York.

Hofstede G. (2001). Culture's consequences: Comparing values, behaviors, institutions, and organizations across nations. Sage. Thousand Oaks, CA.

Homburg C., Jozić D. and Kuehnl C. (2017). Customer experience management: toward implementing an evolving marketing concept. *Journal of the Academy of Marketing Science*, **45**(3), 377–401.

Hood C. (1991). A public management for all seasons? *Public Administration*, **69**(1), 3–19.

Hoske M. (2013). Marketing for Engineers. *Control Engineering*, (April), 8.

Howarth D. and Butler S. (2004). Communicating water conservation: how can the public be engaged? *Water Supply*, **4**(3), 33–44.

Hrudey S. E. and Hrudey E. J. (2004). Safe Drinking Water. Lessons From Recent Outbreaks in Affluent Nations. IWA Publishing, London.

Huberts J., van Eekeren M. and Ernes S. (2000). New marketing tools for water companies. *Water Supply*, **18**(1), 493–496.

Hunt J., Osborn R. and Larson L. (1975). Upper level technical orientation and first level leadership within a noncontingency framework. *Academy of Management Journal*, **18**(3), 476–488.

Hunter G. S. (1990). The water company that turned to money. *Harvard Business Review*, **68**(4), 204.

Hurlimann A. (2007). Recycled water risk perception: a comparison of two case studies. *Water Practice & Technology*, **2**(4), 1–11.

Iacullo R. and Gradilone F. (2000). Customer service benchmarking: a market oriented approach to the delivery of water utility services. *Water Supply*, **18**(1), 500–503.

International Organizational for Standardization (2007). Activities Relating to Drinking Water and Wastewater Services – Guidelines for the Assessment and for the Improvement of the Service to Users, Technical Report ISO 24510: 2007, International Organization for Standardization, Geneva, Switzerland.

Jaiswal A. (2014). How to Reform a Business School the Ivy League Way. Oxford Centre for Higher Education Policy Studies, Oxford, UK.

Jaworski B. J. (2011). On managerial relevance. *Journal of Marketing*, **75**(4), 211–224.

Jehn K. A. and Mannix E. A. (2001). The dynamic nature of conflict: a longitudinal study of intragroup conflict and group performance. *Academy of Management Journal*, **44**(2), 238–251.

Johnson K. and Mortada M. (2005). Effective asset management maximizes your return on investment (ROI) and customer satisfaction. *Water Supply*, **5**(2), 97–106.

Kaplan R. S. and Norton D. P. (1992). The balanced scorecard. Measures that drive performance. *Harvard Business Review*, **70**(1), 71–80.

Kaplan R. S. and Norton D. P. (2004). Strategy Maps: Converting Intangible Assets Into Tangible Outcomes. Harvard Business School Press, Boston.

Karbowiak P. (2003). Marketing a technical business. *Water Supply*, **3**(1–2), 433–440.

Keaveney S. M. (2008). The blame game: an attribution theory approach to marketer–engineer conflict in high-technology companies. *Industrial Marketing Management*, **37**(6), 653–663.

Kemp B., Randle M., Hurlimann A. and Dolničar S. (2012). Community acceptance of recycled water: can we inoculate the public against scare campaigns? *Journal of Public Affairs*, **12**(4), 337–346.

Khastagir A. and Jayasuriya N. (2011). Investment evaluation of rainwater tanks. *Water Resources Management*, **25**(14), 3769–3784.

Kinard B. R. and Capella M. L. (2006). Relationship marketing: the influence of consumer involvement on perceived service benefits. *Journal of Services Marketing*, **20**(6), 359–368.

Klein G., Jiang J. J. and Tesch D. B. (2002). Wanted: project teams with a blend of is professional orientations. *Communications of the ACM*, **45**(6), 81–87.

Korczynski M. (2004). Back-office service work: bureaucracy challenged? *Work, Employment & Society*, **18**(1), 97–114.

Kot M., Castleden H. and Gagnon G. A. (2011). Unintended consequences of regulating drinking water in rural Canadian communities: examples from Atlantic Canada. *Health & Place*, **17**(5), 1030–1037.

Kotler P. (2011). Reinventing marketing to manage the environmental imperative. *Journal of Marketing*, **75**(4), 132–135.

Kotler P. and Levy S. J. (1969). Broadening the concept of marketing. *Journal of Marketing*, **33**(January), 10–15.

Kotler P. and Levy S. J. (1971). Demarketing, yes, demarketing. *Harvard Business Review*, **49**(6), 74–80.

Kotler P. and Zaltman G. (1971). Social marketing: an approach to planned social change. *Journal of Marketing*, **35**(3), 3–12.

Kowalkowski C. (2011). Dynamics of value propositions: insights from service-dominant logic. *European Journal of Marketing*, **45**(1/2), 277–294.

Krishna A. and Morrin M. (2008). Does touch affect taste? The perceptual transfer of product container haptic cues. *Journal of Consumer Research*, **34**(6), 807–818.

Kurland N. B. and Zell D. (2010). Water and business: a taxonomy and review of the research. *Organization & Environment*, **23**(3), 316–353.

Kurland N. B. and Zell D. (2011). Regulating water: a naturological analysis of competing interests among company, town, and state. *Business & Society*, **50**(3), 481–512.

Laing A. (2003). Marketing in the public sector: towards a typology of public services. *Marketing Theory*, **3**(4), 427–445.

Lamb C. W. (1987). Public sector marketing is different. *Business Horizons*, **30**(4), 56–60.

Lamberti L. and Paladino A. (2013). Moving forward with service-dominant logic: exploring the strategic orientations of a service-centred view of the firm. *International Journal of Business Science and Applied Management*, **8**(1), 1–15.

Lancaster G. (1995). Marketing and engineering revisited. *Journal of Business & Industrial Marketing*, **10**(1), 6–15.

Lane R. and Watson M. (2012). Stewardship of things: the radical potential of product stewardship for re-framing responsibilities and relationships to products and materials. *Geoforum*, **43**(6), 1254–1265.

Langfred C. W. (2004). Too much of a good thing? Negative effects of high trust and individual autonomy in self-managing teams. *Academy of Management Journal*, **47**(3), 385–399.

Lansing J. S. (1987). Balinese water temples and the management of irrigation. *American Anthropologist*, **89**(2), 326–341.

Lansing J. S. and Vet T. A. (2012). The functional role of Balinese water temples: a response to critics. *Human Ecology*, **40**(3), 453–467.

Lazarova V. (2013). Global milestones in water reuse: keys to success and trends in development. *Water21*, (August), 12–22.

Lee C. (2011). Privatization, water access and affordability: evidence from Malaysian household expenditure data. *Economic Modelling*, **28**(5), 2121–2128.

Lee N. and Lings I. (2008). Doing Business Research: A Guide to Theory and Practice. SAGE, London.

Leonard L. (2013). The water crisis in ireland. the sociopolitical contexts of risk in contemporary society. In: The Social Life of Water, J. R. Wagner (ed.), Berghahn Books, New York.

Levallois P., Grondin J. and Gingras S. (1999). Evaluation of consumer attitudes on taste and tap water alternatives in Quebec? *Water Science and Technology*, **40**(6), 135–140.

Levitt T. (1960). Marketing Myopia. *Harvard Business Review*, **38**(May/June), 45–56.

Levy D. (1982). Diamonds, water, and Z goods: an account of the paradox of value. *History of Political Economy*, **14**(3), 312–322.

Lewin K. (1951). Field Theory in Social Science: Selected Theoretical Papers. Harper & Row, New York, NY.

Lin C. and Berg S. V. (2008). Incorporating service quality into yardstick regulation: an application to the Peru water sector. *Review of Industrial Organization*, **32**(1), 53–75.

Lings I. N. and Greenley G. E. (2010). Internal market orientation and market-oriented behaviours. *Journal of Service Management*, **21**(3), 321–343.

Loh L., Sankar C. S. and Yeong W. Y. (1995). Job orientation, perceptions, and satisfaction a study of information technology professionals in Singapore. *Information & Management*, **29**(5), 239–250.

Lovelock C. H. (1983). Classifying services to gain strategic marketing insights. *Journal of Marketing*, **47**(Summer), 9–20.

Lovelock C. H. (1992). Cultivating the flower of service: new ways of looking at core and supplementary services. In: Marketing, Operations, and Human Resources: Insights Into Services, P. Eiglier and E. Langeard (eds), IAE, Université d'Aix-Marseille III, Aix-en-Provence, France, pp. 296–316.

Lovelock C. H., Patterson P. G. and Walker R. H. (2007). Services Marketing: An Asia-Pacific and Australian perspective, 4th edn., Pearson Education Australia, Frenchs Forest NSW.

Lowe B., Lynch D. and Lowe J. (2014). The role and application of social marketing in managing water consumption: a case study: social marketing and water consumption. *International Journal of Nonprofit and Voluntary Sector Marketing*, **19**(1), 14–26.

Lusch R. F. and Vargo S. L. (2014). Service-Dominant Logic: Premises, Perspectives, Possibilities. Cambridge University Press, Cambridge.

Macdonald E. K., Wilson H., Martinez V. and Toossi A. (2011). Assessing value-in-use: a conceptual framework and exploratory study. *Industrial Marketing Management*, **40**(5), 671–682.

MacGillivray B. H. (2014). Fast and frugal crisis management: an analysis of rule-based judgment and choice during water contamination events. *Journal of Business Research*, **67**(8), 1717–1724.

Macomber J. D. (2013). Building sustainable cities. *Harvard Business Review*, **91**(7/8), 40–50.

Madill J. and Ziegler R. (2012). Marketing social missions-adopting social marketing for social entrepreneurship? A conceptual analysis and case study: marketing social missions. *International Journal of Nonprofit and Voluntary Sector Marketing*, **17**(4), 341–351.

Mahood Q., Eerd D. V. and Irvin E. (2014). Searching for grey literature for systematic reviews: challenges and benefits. *Research Synthesis Methods*, **5**(3), 221–234.

Marchesan M. and Morran J. (2004). Tastes associated with products in contact with drinking water. *Water Science & Technology*, **49**(9), 227–231.

Marcuse H. (1964). One Dimensional Man. Abacus, London.

Marketline (2015). Global Water Utilities, Technical Report 0199-2153, Datamonitor, London.

Maroto M. L. (2015). Pathways into bankruptcy: accumulating disadvantage and the consequences of adverse life events. *Sociological Inquiry*, **85**(2), 183–216.

Maslow A. H. (1943). A theory of human motivation. *Psychological Review*, **50**(4), 370–396.

McCarthy J. E. (1964). Basic Marketing. A Managerial Approach. Irwin, Homewood, IL.

McCoy M. (2014). Reputational threat and image repair strategies: Northern Ireland water's crisis communication in a freeze/thaw incident. *Journal of Nonprofit & Public Sector Marketing*, **26**(2), 99–126.

McGuire M. J. (1995). Off-flavor as the consumer's measure of drinking water safety. *Water Science and Technology*, **31**(11), 1–8.

Means E., Chowdhury Z., Westerhoff G., Passantino L. and Ruettan J. (2008). Communicating the Value of Water: An Introductory Guide for Water Utilities. AWWA Research Foundation, Denver.

Meij S., Ruiters C. and Stumphius C. (2005). Market-driven pricing structures for drinking water. *Water Supply*, **5**(6), 225–233.

Meissner R. (2015). The relevance of social theory in the practice of environmental management. *Science and Engineering Ethics*, **22**(5), 1345–1360.

Mendoza R. L. (2011). Fluoride-treated water and the problem of merit goods. *Water Policy*, **13**(1), 38.

Menegaki A. N., Mellon R. C., Vrentzou A., Koumakis G. and Tsagarakis K. P. (2009). What's in a name: framing treated wastewater as recycled water increases willingness to use and willingness to pay. *Journal of Economic Psychology*, **30**(3), 285–292.

Merkel W., Staben N., Bräcker J., Levai P., Neskovic M., Petry D. and Weiss M. (2011). Structural conditions in German water supply. *Water Science & Technology: Water Supply*, **11**(5), 596.

Mintzberg H. (2005). Managers, Not MBAs: A Hard Look at the Soft Practice of Managing and Management Development. Berrett-Koehler. London, San Francisco.

Mitchell C. (2002). Selling the Brand Inside. *Harvard Business Review*, **80**(1), 99–101, 103–5, 126.

Morgan N. (1990). Professional accountancy firms and marketing. *Service Industries Journal*, **10**(3), 599–613.

Morgan R. A. and Smith J. L. (2013). Premodern streams of thought in twenty–first–century water management. *Radical History Review*, **2013**(116), 105–129.

Mugabi J. and Njiru C. (2006). Marketing urban water services: information needs for water utility managers in developing countries. *Water Science & Technology: Water Supply*, **6**(5), 17.

Mugisha S. (2007). Performance assessment and monitoring of water infrastructure: an empirical case study of benchmarking in Uganda. *Water Policy*, **9**(5), 575–491.

Mukokoma M. M. N. and Van Dijk M. P. (2013). New public management reforms and efficiency in urban water service delivery in developing countries: blessing or fad? *Public Works Management & Policy*, **18**(1), 23–40.
Nagata J. M., Valeggia C. R., Smith N. W., Barg F. K., Guidera M. and Bream K. D. W. (2011). Criticisms of chlorination: social determinants of drinking water beliefs and practices among the Tz'utujil Maya. *Rev Panam Salud Publica*, **29**(1), 9–16.
Naipaul S. and Parsa H. G. (2000). Supplementary services as a differentiation strategy. *Journal of Quality Assurance in Hospitality & Tourism*, **1**(1), 67–80.
Narver J. C. and Slater S. F. (1990). The effect of a market orientation on business profitability. *Journal of Marketing*, **54**(4), 20–35.
National Health and Medical Research Council (2011). Australian Drinking Water Guidelines. Canberra.
Nelson C. and Peterson S. R. (1982). If you're an engineer, you're probably a utilitarian. *Issues in Engineering: Journal of Professional Activities*, **108**(1), 13–18.
Newman M. E. J. and Girvan M. (2004). Finding and evaluating community structure in networks. *Physical Review E*, **69**(2), 026113-1–026113-15.
Nordhaus W. D. (1997). Do real-outcome and real-wage measures capture reality? The history of lighting suggests not. In: The Economy of New Goods, T. F. Bresnahan and R. J. Gordon (eds), University of Chicago Press, Chicago, pp. 29–66.
Norman C. (1997). Wanted: marketing engineers. *Appliance Manufacturer*, **45**(7), 49–51.
Osborne S. P., Radnor Z. and Nasi G. (2013). A New Theory for Public Service Management? Toward a (Public) Service-Dominant Approach. *The American Review of Public Administration*, **43**(2), 135–158.
Osterwalder A., Pigneur Y., Bernarda G., Smith A. and Papadakos P. (2014). Value Proposition Design: How to Create Products and Services Customers Want. Get Started with Strategyzer Series. Wiley, Hoboken, NJ.
Ostrom E. (1990). Governing the Commons: The Evolution of Institutions for Collective Action. The Political Economy of Institutions and Decisions. Cambridge University Press, Cambridge, New York.
Ostrom E. (1999). Revisiting the commons: local lessons, global challenges. *Science*, **284**(5412), 278–282.
Parasuraman A., Zeithaml V. and Berry L. L. (1985). A conceptual model of service quality and its implications for future research. *Journal of Marketing*, **49**(4), 41–50.
Parasuraman A., Zeithaml V. and Berry L. L. (1988). SERVQUAL: a multiple-item scale for measuring consumer perceptions of service quality. *Journal of Retailing*, **64**(1), 12–40.
Parker D. (2012). The Official History of Privatisation. Taylor & Francis, Hoboken, Vol. II.
Parr J. (2005). Local water diversely known: Walkerton Ontario, 2000 and after. *Environment and Planning D: Society and Space*, **23**(2), 251–271.
Parry M. E. and Song X. M. (1993). Determinants of R&D-marketing integration in high-tech Japanese firms. *Journal of Product Innovation Management*, **10**(1), 4–22.
Pasquier M. (2012). Public sector marketing. In: Encyclopedic Dictionary of Public Administration, L. Côté and F. Savard (eds), Administration, page n.p. École nationale d'administration publique, Quebec City.
Patsiaouras G., Saren M. and Fitchett J. A. (2015). The marketplace of life? An exploratory study of the commercialization of water resources through the lens of macromarketing. *Journal of Macromarketing*.

Payne A. and Frow P. (2014). Deconstructing the value proposition of an innovation exemplar. *European Journal of Marketing*, **48**(1/2), 237–270.

Payne A., Frow P. and Eggert A. (2017). The customer value proposition: evolution, development, and application in marketing. *Journal of the Academy of Marketing Science*, **45**(4), 467–489.

Phipps M. and Brace-Govan J. (2011). From right to responsibility: sustainable change in water consumption. *Journal of Public Policy and Marketing*, **30**(2), 203–219.

Plangger K., Kietzmann J. H., Pitt L. F., Berthon P. and Hannah D. (2013). Nomen est omen: formalizing customer labeling theory. *AMS Review*, **3**(4), 193–204.

Porter M. (1979). How competitive forces shape strategy. *Harvard Business Review*, **57**(2), 137–145.

Prevos P. (2013). ServAqua: Towards a Model for Service Quality in Potable Reticulated Water Services. Academy of Marketing Science, Melbourne.

Prevos P. (2015). Visualising water quality: a graphical index for drinking water system performance. In: OzWater, Australian Water Association, Adelaide.

Prevos P. (2016). The Invisible Water Utility. Employee Behaviour and Customer Experience in Service-Dominant Logic. PhD thesis, La Trobe University, Melbourne, Australia.

Prüss-Üstün A., Bos R., Gore F. and Bartram J. (2008). Safer Water, Better Health. World Health Organization, Geneva.

Psaila A. (2017). A Service Quality Model for the Local Water Utility: A Customer Perspective. MBA, University of Malta, Malta.

Puffer M. and Al-Musallam L. B. A. (2007). Water stewardship in the desert: water conservation projects in the kingdom of Saudi Arabia. *Water Practice & Technology*, **2**(4), wpt2007074.

Puget S., Beno N., Chabanet C., Guichard E. and Thomas-Danguin T. (2010). Tap water consumers differ from non-consumers in chlorine flavor acceptability but not sensitivity. *Water Research*, **44**(3), 956–964.

Rambo E., Baumgartner R. and Koenig C. (2003). Developing Customer Service Targets by Assessing Customer Perspectives. AWWA Research Foundation & International Water Association, London.

Rangan V. K., Chu M. and Petkoski D. (2011). Segmenting the base of the pyramid. *Harvard Business Review*, **89**(6), 113–117.

Raucher R., Clements J., Oxenford J., Raucher K. and McLoughlin B. 2012). Expanding Water Utility Services Beyond Water Supply, Technical Report 4171.

Reed B. and Reed B. (2011). How Much Water is Needed in Emergencies? Technical Report 9, World Health Organisation, Geneva.

Reibstein D. J., Day G. and Wind J. (2009). Guest editorial: is marketing academia losing its way? *Journal of Marketing*, **73**(4), 1–3.

Reid S., Chalak A. and Hecht J. (2010). Determining the optimal investment plan for water utilities: the case of Veolia Water Central. *Water Science & Technology: Water Supply*, **10**(3), 367.

Renaghan L. M. (1981). A new marketing mix for the hospitality industry. *Cornell Hotel and Restaurant Administration Quarterly*, **22**(2), 31–36.

Revella A. (2015). Buyer Personas: How to Gain Insight Into Your Customer's Expectations, Align Your Marketing Strategies, and Win More Business. The new rules series. Wiley, Hoboken, NJ. OCLC: 931760250.

Rexha N., Kirk-Burnnand T. and Ramaseshan B. (2000). The impact of technically focused values on market orientation of professional service firms. *Journal of Professional Services Marketing*, **21**(1), 45–61.

Richter C. P. (2010). Automatic dishwashers: efficient machines or less efficient consumer habits? *International Journal of Consumer Studies*, **34**(2), 228–234.

Richter C. P. (2011). Usage of dishwashers: observation of consumer habits in the domestic environment: observation of dishwasher use. *International Journal of Consumer Studies*, **35**(2), 180–186.

Ridolfi E., Müller N. and Hoffjan A. (2012). Customer Perspectives on New Urban Water Services—Literature Review, Technical Report Task 2.2.2.

Roa D. V., O'Callaghan P. and Value of Water Coalition (2014). *The value of water: a compendium of essays by smart CEOs*.

Robertson H. M. and Taylor W. L. (1957). Adam Smith's approach to the theory of value. *The Economic Journal*, **67**(266), 181–198.

Robson A. (2007). A 'public good' is not just something which is 'good for the public'. The abuse of economic terminology in public debate. *Review*, (July), 39–40.

Robyns B., Davigny A., Francois B., Henneton A. and Sprooten J. (2012). Electricity Production from Renewables Energies. Wiley, Hoboken.

Roggeveen A. L., Tsiros M. and Grewal D. (2012). Understanding the co-creation effect: when does collaborating with customers provide a lift to service recovery? *Journal of the Academy of Marketing Science*, **40**(6), 771–790.

Rook D. W. (1985). The ritual dimension of consumer behavior. *Journal of Consumer Research*, **12**(3), 251–264.

Rosenbaum M. S., Otalora M. L. and Ramírez G. C. (2017). How to create a realistic customer journey map. *Business Horizons*, **60**(1), 143–150.

Ruas Dos Santos, M. (2000). How is water perceived: as a commodity or as a product? *Water Supply*, **18**(1), 504–508.

Rushton A. M. and Carson D. J. (1985). The marketing of services: managing the intangibles. *European Journal of Marketing*, **19**(3), 19–40.

Russell G. (2009). Water for Gold!: The Fight to Quench Central Victoria's Goldfields. Australian Scholarly Pub, North Melbourne, Vic.

Rust R. T. and Oliver R. L. (eds) (1994). Service Quality: New Directions in Theory and Practice. Sage Publications, Thousand Oaks, London & New Delhi.

Ryan B. (1991). Circulation matters. Towards a model of public sector marketing. In: Service Delivery and Public Sector Marketing, C. O'Faircheallaigh, J. Wanna and P. M. Weller (eds), Macmillan Co. of Australia, South Melbourne.

Salzman J. (2012). Drinking Water: A History. Duckworth Overlook, London.

Schmenner R. W. (1986). How can service businesses survive and prosper? *Sloan Management Review*, **27**(3), 21–32.

Schnoor J. L. (2011). Water–energy nexus. *Environmental Science & Technology*, **45**(12), 5065–5065.

Selinger C. (2004). Stuff You Don't Learn in Engineering School: Skills for Success in the Real World. IEEE Press; Wiley-Interscience, Piscataway, NJ: Hoboken NJ.

Sharp L. (2006). Water demand management in England and Wales: constructions of the domestic water user. *Journal of Environmental Planning and Management*, **49**(6), 869–889.

Shaw C. T., Shaw V. and Enke M. (2004). Relationships between engineers and marketers within new product development: an Anglo-German comparison. *European Journal of Marketing*, **38**(5/6), 694–719.

Shaw V. and Shaw C. T. (1998). Conflict between engineers and marketers. *Industrial Marketing Management*, **27**(4), 279–291.

Shaw V. and Shaw C. T. (2003). Marketing: the Engineer's perspective. *Journal of Marketing Management*, **19**(3/4), 345–378.

Shostack G. (1977). Breaking free from product marketing. *Journal of Marketing*, **41**(2), 73.

Shostack G. (1982). How to design a service. *European Journal of Marketing*, **16**(1), 49–63.

Shove E. (2003). Converging conventions of comfort, cleanliness and convenience. *Journal of Consumer Policy*, **26**(4), 395–418.

Skellett C. (1995). Understanding and meeting the needs of our customers. *Managing Service Quality*, **5**(4), 22–24.

Smith A. (2010). The Wealth of Nations. Capstone, Chichester.

Sofoulis Z. (2005). Big water, everyday water: a sociotechnical perspective. *Continuum: Journal of Media & Cultural Studies*, **19**(4), 445–463.

Sofoulis Z. (2010). Water Managers' Views on the Social Dimensions of Urban Water, Technical Report Report from Cross-Connections: Linking Urban Water Managers with Humanities, Arts and Social Sciences Researchers, National Water Commission, Canberra.

Solomon M. R., Russell-Bennett R. and Previte J. (2010). Consumer Behaviour: Buying, Having, Selling. Pearson, Frenchs Forest, NSW.

Song M. and Thieme R. J. (2006). A cross-national investigation of the R&D–marketing interface in the product innovation process. *Industrial Marketing Management*, **35**(3), 308–322.

Soutar G. (2013). Service quality, customer satisfaction, and value: An examination of their relationships. In: Service Quality Management in Hospitality, Tourism and Leisure, J. Kadampully, C. Mok and B. Sparks (eds), Taylor and Francis, New York and London, pp. 97–110.

Steenkamp J. (1989). Product Quality: An Investigation into the Concept and How It Is Perceived by Consumers. Van Gorcum, Assen/Maastricht, the Netherlands.

Stell R. and Donoho C. L. (1996). Classifying services from a consumer perspective. *Journal of Services Marketing*, **10**(6), 33–44.

Stevenson E. G., Greene L. E., Maes K. C., Ambelu A., Tesfaye Y. A., Rheingans R. and Hadley C. (2012). Water insecurity in 3 dimensions: an anthropological perspective on water and women's psychosocial distress in Ethiopia. *Social Science & Medicine*, **75**(2), 392–400.

Stevenson S. (2006). The abnormal performance of UK utility privatisations. *Studies in Economics and Finance*, **23**(3), 164.

Strandvik T., Holmlund M. and Grönroos C. (2014). The mental footprint of marketing in the boardroom. *Journal of Service Management*, **25**(2), 241–252.

Strang V. (2004). The Meaning of Water. Berg, Oxford; New York.

Suffet I. M., Khiari D. and Bruchet A. (1999). The drinking water taste and odor wheel for the millennium: beyond geosmin and 2-methylisoborneol. *Water Science and Technology*, **40**(6), 1–13.

Svensson G. (2005). The spherical marketing concept: a revitalization of the marketing concept. *European Journal of Marketing*, **39**(1/2), 5–15.

Syed T. H., Akhtar N. and Khan N. (2011). Consumer or customer?. Does the literature care for a difference? *Interdisciplinary Journal of Contemporary Research in Business*, **2**(12), 618–628.

Tan J. (2012). The pitfalls of water privatization: failure and reform in Malaysia. *World Development*, **40**(12), 2552–2563.

Tanji H., Kiri H. and Iida T. (2012). Service Dominant Logic for Irrigation Services. Adelaide.

Tarus B. K. (2014). Effects of job rotation strategy on high performance workplace, in Lake Victoria North water services board, Kenya. *International Journal of Business and Management*, **9**(11), 139–146.

Taylor M. (2011). Engineers and marketing people must be friends. *Professional Engineering*, **24**(11), 8.

Teillet E., Schlich P., Urbano C., Cordelle S. and Guichard E. (2010). Sensory methodologies and the taste of water. *Food Quality and Preference*, **21**(8), 967–976.

Ten Bos, R. (2014). Water. Een Geofilosofische Geschiedenis [Water: A Geophilosophical History]. Boom, Amsterdam.

Thomas D. R. (1978). Strategy is different in service businesses. *Harvard Business Review*, **56**(4), 158–165.

Thomson P. J. (2013). Value Proposition Canvas Template.

Tian R. G. (2013). Business anthropology: an overview. In: Rethinking Business Anthropology: Cultural Strategies in Marketing and Management, A. H. Walle (ed.) Greenleaf, Sheffield.

TMI/SOCAP (2005). Complaint Culture Survey National Report.

Tomar A. and Dhiman A. (2012). Customer oriented bureaucracy & flexibility: interactions & implications in the services industry context. *Indian Journal of Industrial Relations*, **48**(1), 43–58.

Trevelyan J. P. (2014). *The Making of an Expert Engineer*.

Tsindos S. (2012). What drove us to drink two litres of water a day? *Australian and New Zealand Journal of Public Health*, **36**(3), 205–207.

Tupper H. and Resende M. (2004). Efficiency and regulatory issues in the Brazilian water and sewage sector: an empirical study. *Utilities Policy*, **12**(1), 29–40.

United Nations (2012). System of Environmental-Economic Accounting for Water. Economic & Social Affairs. United Nations, New York.

van den Berg C. (1997). Water privatization and regulation in England and Wales. *Public Policy for the Private Sector*, (115).

van der Valk W. and Axelsson B. (2015). Towards a managerially useful approach to classifying services. *Journal of Purchasing and Supply Management*, **21**(2), 113–124.

Vannini P. and Taggart J. (2015). Off the Grid: Re-Assembling Domestic Life. Innovative Ethnographies. Routledge, New York.

Van Vliet B., Chappells H. and Shove E. (2005). Infrastructures of Consumption Environmental Innovation in the Utility Industries. Earthscan, London; Sterling, VA.

Van Waterschoot W., Van Den Bulte, C. (1992). The Four-P classification of the marketing mix revisited. *Journal of Marketing*, **56**(4), 83–93.

Vargo S. L. and Lusch R. F. (2004). Evolving to a new dominant logic for marketing. *Journal of Marketing*, **68**(1), 1–17.

Vargo S. L. and Lusch R. F. (2008). Service-dominant logic: continuing the evolution. *Journal of the Academy of Marketing Science*, **36**(1), 1–10.

Vargo S. L. and Lusch R. F. (2016). Institutions and axioms: an extension and update of Service-Dominant Logic. *Journal of the Academy of Marketing Science*, **44**(1), 5–23.

Vining A. R. (2011). Public agency external analysis using a modified "five forces" framework. *International Public Management Journal*, **14**(1), 63–105.

Visser J. (1996). Marketing for engineers. *New Zealand Engineering*, **51**(2), 24.

Vloerbergh I., Fife-Schaw C., Kelay T., Jonathan Chenoweth Morrison G. and Lundéhn C. (2007). Assessing Consumer Preferences for Drinking Water Services. Methods for Water Utilities. Techneau, n.p.

Wagner J. R. (ed.) (2013). The Social Life of Water. Berghahn Books, New York.

Walker R. H. and Johnson L. W. (2009). Signalling intrinsic service quality and value via accreditation and certification. *Managing Service Quality*, **19**(1), 85–105.

Walker R. H., Hanson D., Nelson L. and Fisher C. (1998). A case for more integrative multi-disciplinary marketing education. *European Journal of Marketing*, **32**(9/10), 803–812.

Walker R. H., Johnson L. W. and Leonard S. (2006). Re-thinking the conceptualization of customer value and service quality within the service-profit chain. *Managing Service Quality*, **16**(1), 23–36.

Walsh K. (1994). Marketing and public sector management. *European Journal of Marketing*, **28**(3), 63–71.

Walton A. and Hume M. (2011). Creating positive habits in water conservation: the case of the Queensland Water Commission and the Target 140 campaign. *International Journal of Nonprofit and Voluntary Sector Marketing*, **16**(3), 215–224.

Wasserman S. and Faust K. (1994). Social Network Analysis: Methods and Applications. Cambridge University Press, Cambridge.

Watson A., Viney H. and Schomaker P. (2002). Consumer attitudes to utility products: a consumer behaviour perspective. *Marketing Intelligence & Planning*, **20**(7), 394–404.

Wayman E. R. (2008). Consumer Anthropology. *Current Anthropology*, **49**(2), 169–169.

Webber M. A., Atherton P. and Newcombe G. (2015). Taste and odour and public perceptions: what do our customers really think about their drinking water? *Journal of Water Supply: Research and Technology – Aqua*, **64**(7), 802–811.

Weidner K. L., Rosa J. A. and Viswanathan M. (2010). Marketing to subsistence consumers: lessons from practice. *Journal of Business Research*, **63**(6), 559–569.

Whelan S., Davies G., Walsh M. and Bourke R. (2010). Public sector corporate branding and customer orientation. *Journal of Business Research*, **63**(11), 1164–1171.

Wickham M. (2009). Thana-marketing strategy: exploring the 8 Ps that dare not speak their name. *International Journal of Business Strategy*, **9**(1), 194–201.

Widrick S. M. (1985). Quantity surcharge—quantity discount: pricing as it relates to quantity purchased. *Business & Society*, **24**(1), 1–7.

Wijkman A. and Gordon L. (2002). Workshop 9 (synthesis): how to increase the status of water issues in governance and in public perception. *Water Science & Technology*, **45**(8), 229–231.

Williams R. F. and Mountjoy N. (2012). Seeing water in a spatial and temporal perspective: The residential demand for water in regional Victoria. Melbourne, Australia. School of Economics, La Trobe University, Bendigo, Australia.

Winick C. (1961). Anthropology's contributions to marketing. *Journal of Marketing*, **25**(July), 53–60.

Wirtz J. (1998). Development of a Service Guarantee Model. *Asia Pacific Journal of Management*, **15**(1), 51–75.

Workman J. (1995). Engineering's interactions with marketing groups in an engineering-driven organization. *IEEE Transactions on Engineering Management*, **42**(2), 129–139.

Workman J. P. (1993). Marketing's limited role in new product development in one computer systems firm. *Journal of Marketing Research (JMR)*, **30**(4), 405–421.

Zaichkowsky J. L. (1985). Measuring the involvement construct. *Journal of Consumer Research*, **12**(3), 341–352.

Zaichkowsky J. L. (1994). The personal involvement inventory: reduction, revision, and application to advertising. *Journal of Advertising*, **23**(4), 59.

Zetland D. (2008). Conflict and cooperation within an organization: a case study of the metropolitan water district of Southern California. *SSRN Electronic Journal*.

Zetland D. (2011). The End of Abundance: Economic Solutions to Water Scarcity. Aguanomics Press, Amsterdam and Mission Viejo.

Zubizarreta J. (2008). Drought management and customer service. *Water Practice & Technology*, **3**(1), wpt2008015.

Index

Appendix A

IWA Journal Abstract List

The abstracts of the articles listed below were used in content analysis (Chapter 2). Any articles which where sourced in full are also listed in the References section.

Abbaszadegan M., Gerba C. P., Yahya M. T. and Rusin P. (1993). Evaluation of a microbiological water purifier for inactivation of viruses, bacteria and Giardia cysts. *Water Science & Technology*, **27**(3–4), 329–333.

Adams W. B. (2008). Major water Industry reform in Australia: a case study from Queensland. *Water Practice & Technology*, **3**(4), 1–11.

Addo-Yobo F. N. and Njiru C. (2006). Role of consumer behaviour studies in improving water supply delivery to the urban poor. *Water Policy*, **8**(2), 111–126.

Addo-Yobo F. N., Njiru C. and Sohail M. (2006). Determinants of households' intention to pay for improved water services: an application of the theory of reasoned action. *Journal of Water Supply: Research and Technology—Aqua*, **55**(6), 419–425.

Aguilar-Benitez I. and Saphores J.-D. (2010). Public accountability and performance of two border water utilities. *Water Policy*, **12**(2), 203–219.

Akpinar M. G. and Gul M. (2014). An assessment of consumer preferences on the drinking water market: today to the future. *Journal of Water Supply: Research and Technology—Aqua*, **63**(7), 525–531.

Alhumoud J. M. (2008). Freshwater consumption in Kuwait: analysis and forecasting. *Journal of Water Supply: Research and Technology—Aqua*, **57**(4), 279.

Allan R., Jeffrey P., Clarke M. and Pollard S. (2013). The impact of regulation, ownership and business culture on managing corporate risk within the water industry. *Water Policy*, **15**(3), 458.

Marketing Urban Water Supply
Peter Prevos
doi: 10.2166/9781780408675_183

Ameyaw E. E., Memon F. A. and Bicik J. (2013). Improving equity in intermittent water supply systems. *Journal of Water Supply: Research and Technology—Aqua*, **62**(8), 552–562.

Amoatey P. K., Minke R. and Steinmetz H. (2013). Leakage estimation in water networks based on two categories of night-time users: a case study of a developing country network. *Water Science & Technology: Water Supply*, **14**(2), 329–336.

Arbon M. and Ireland M. (2003). Water recycling: a major new initiative for Melbourne—crucial for a sustainable future. *Water Science & Technology*, **47**(7–8), 57–63.

Archer A. R., Elmore A. C., Bell E. and Rozycki C. (2011). Field investigation of arsenic in ceramic pot filter-treated drinking water. *Water Science & Technology*, **63**(10), 2193–2198.

Arvin E., Engelsen P. and Sebber U. (2005). Biodegradation of gasoline compounds (BTEX) in a water works sand filter. *Water Supply*, **4**(5–6), 29–33.

Bae B. U., Shin H. S. and Choi J. J. (2007). Taste and odour issues in South Korea's drinking water industry. *Water Science & Technology*, **55**(5), 203.

Bakker M., Trietsch E. A., Vreeburg J. H. G. and Rietveld L. C. (2014). Analysis of historic bursts and burst detection in water supply areas of different size. *Water Science & Technology: Water Supply*, **14**(6), 1035–1044.

Banerjee A., McFarland D. A., Singh R. and Quick R. (2007). Cost and financial sustainability of a household-based water treatment and storage intervention in Zambia. *Journal of Water and Health*, **5**(3), 385–394.

Baron J. (2001). Monitoring strategy for lead in drinking water at consumer's tap: field experiments in France. *Water Science & Technology: Water Supply*, **1**(4), 193–200.

Barrett G. and Wallace M. (2009). Characteristics of Australian urban residential water users: implications for water demand management and whole of the system water accounting framework. *Water Policy*, **11**(4), 413–426.

Baudin I., Cagnard O., Grandguillaume J. J. and Do-Quang Z. (2006). Algae and associated toxins and metabolites: methodology for risk assessment and risk management. *Water Practice and Technology*, **1**(4), wpt2006081–wpt2006081.

Baur R., Gauffre P. L. and Saegrov S. (2003). Multi-criteria decision support for annual rehabilitation programmes in drinking water networks. *Water Supply*, **3**(1–2), 43–50.

Berney M., Vital M., Hülshoff I., Weilenmann H.-U., Egli T. and Hammes F. (2008). Rapid, cultivation-independent assessment of microbial viability in drinking water. *Water Research*, **42**(14), 4010–4018.

Besner M.-C., Prévost M. and Regli S. (2011). Assessing the public health risk of microbial intrusion events in distribution systems: conceptual model, available data, and challenges. *Water Research*, **45**(3), 961–979.

Bichai F. and Smeets P. W. M. H. (2013). Using QMRA-based regulation as a water quality management tool in the water security challenge: experience from the Netherlands and Australia. *Water Research*, **47**(20), 7315–7326.

Bicik J., Kapelan Z., Makropoulos C. and Savic D. A. (2011). Pipe burst diagnostics using evidence theory. *Journal of Hydroinformatics*, **13**(4), 596–608.

Blette V. (2008). Drinking water public right-to-know requirements in the United States. *Journal of Water and Health*, **6**(S1), s43–s51.

Boxall J. B. and Prince R. A. (2006). Modelling discolouration in a Melbourne (Australia) potable water distribution system. *Journal of Water Supply: Research and Technology—Aqua*, **55**(3), 207–219.

Boxall J. B., Skipworth P. J. and Saul A. J. (2003). Aggressive flushing for discolouration event mitigation in water distribution networks. *Water Supply*, **3**(1–2), 179–186.

Bradshaw R. A., Illaszewicz G. L. and Avrahamson Y. G. (2013). Introducing layer of protection analysis for water safety risk assessments. *Water Quality Research Journal of Canada*, **48**(1), 76–84.

Breach R. A. (1999). Overview of residual disinfection practice by European water suppliers. *Journal of Water Supply: Research and Technology—Aqua*, **48**(2), 39–43.

Breitenmoser A., Fretz R., Schmid J., Besl A. and Etter R. (2011). Outbreak of acute gastroenteritis due to a washwater-contaminated water supply, Switzerland, 2008. *Journal of Water and Health*, **9**(3), 569–576.

Bruvold W. H. (1970). Mineral taste and the potability of domestic water. *Water Research*, **4**(5), 331–340.

Buckle J. S. (2005). Social interaction in water demand management implementation. *Water Supply*, **5**(3–4), 295–301.

Burgess M. D. (1991). Planning for reuse. Developing a strategy for the Northern Territory, Australia. *Water Science & Technology*, **24**(9), 31–43.

Burlingame G. A. and Mackey E. D. (2007). Philadelphia obtains useful information from its customers about taste and odour quality. *Water Science & Technology*, **55**(5), 257.

Burn S., Tucker S., Rahilly M., Davis P., Jarrett R. and Po M. (2003). Asset planning for water reticulation systems: the PARMS model. *Water Supply*, **3**(1–2), 55–62.

Burn S., Marlow D. and Tran D. (2010). Modelling asset lifetimes and their role in asset management. *Journal of Water Supply: Research and Technology—Aqua*, **59**(6–7), 362.

Campos C. and Harmant P. (2002). Assessing the impact of dissolved organic carbon changes on disinfectant stability in a distribution system. *Water Supply*, **2**(3), 251–257.

Camps L. M. (2000). Quality of service and customer satisfaction. *Water Supply*, **18**(1), 497–499.

Cardew P. T. (2000). Simulation of lead compliance data. *Water Research*, **34**(8), 2241–2252.

Carpenter G. (2003). Making markets work for clean water. *Water Science & Technology*, **47**(6), 97–102.

Cembrano G., Quevedo J., Puig V., Pérez R., Figueras J., Verdejo J. M. and Casas M. (2011). PLIO: a generic tool for real-time operational predictive optimal control of water networks. *Water Science & Technology*, **64**(2), 448–459.

Christensen S. C. B., Nissen E., Arvin E. and Albrechtsen H.-J. (2012). Influence of asellus Aquaticus on escherichia coli, klebsiella pneumoniae, campylobacter jejuni and naturally occurring heterotrophic bacteria in drinking water. *Water Research*, **46**(16), 5279–5286.

Clement M., Seux R. and Rabarot S. (2000). A practical model for estimating total lead intake from drinking water. *Water Research*, **34**(5), 1533–1542.

Coelho S. T., James S., Sunna N., Abu Jaish A. and Chatila J. (2003). Controlling water quality in intermittent supply systems. *Water Research*, **3**(1–2), 119–125.

Cole G., O'Halloran K. and Stewart R. A. (2012). Time of use tariffs: implications for water efficiency. *Water Science & Technology: Water Supply*, **12**(1), 90.

Cook D. and Newcombe G. (2004). Can we predict the removal of MIB and geosmin with PAC by using water quality parameters? *Water Supply*, **4**(4), 221–226.

Cook D., Newcombe G. and Sztajnbok P. (2001). The application of powdered activated carbon for MIB and geosmin removal: predicting PAC doses in four raw waters. *Water Research*, **35**(5), 1325–1333.

Cooper R. (2000). Creative use of information technology WISP—Water Industry Service Provision. *Water Supply*, **18**(1), 509–513.

Corbi E., Valérie J., Alain Q., Francine M., Adeline L. and Franck T. (2009). Understanding and forecasting taste-and-odour formation in a drinking water distribution system: case study of Lyon (France). *Water Science & Technology: Water Supply*, **9**(1), 67–74.

Crampton A. and Ragusa A. T. (2014). Perceived agricultural runoff impact on drinking water. *Journal of Water and Health*, **12**(3), 484–491.

Croke G. and Williams J. (2011). Customer consultation in water project planning and delivery: lessons from case studies. *Water Practice & Technology*, **5**(4), 1–8.

Cross P. and Morel A. (2005). Pro-poor strategies for urban water supply and sanitation services delivery in Africa. *Water Science & Technology*, **51**(8), 51–57.

de Greef E., Zoeteman B. C. J., Oers H. J. van, Kóster E. P. and Rook J. J. (1983). Drinking water contamination and taste assessment by large consumer panels. *Water Science & Technology*, **15**(6–7), 13–24.

Diás S. L., Lucas H., do Céu Almeida, M., Alegre H., Vriato M. and Loureiro D. (2007). Identification of opportunities to improve efficiency by water consumption assessment. *Water Science & Technology: Water Supply*, **7**(4), 57–63.

Dietrich A. M., Whelton A. J., Hoehn R. C. and Anderson R. (2004). The attribute rating test for sensory analysis. *Water Science & Technology*, **49**(9), 61–67.

Dietrich A. M., Phetxumphou K. and Gallagher D. L. (2014). Systematic tracking, visualizing, and interpreting of consumer feedback for drinking water quality. *Water Research*, **66**, 63–74.

Dizer H., Dürkop J., Grohmann A., Kopecka H. and López-Pila J. M. (1993). Viruses and water: problems, detection, and control. *Water Science & Technology*, **27**(7–8), 127–133.

Dolničar S. and Hurlimann A. (2009). Drinking water from alternative water sources: differences in beliefs, social norms and factors of perceived behavioural control across eight Australian locations. *Water Science & Technology*, **60**(6), 1433.

Dolničar S., Hurlimann A. and Grún B. (2014). Branding water. *Water Research*, **57**, 325–338.

Doria M. de F. (2006). Bottled water versus tap water: understanding consumers-preferences. *Journal of Water and Health*, **4**(2), 271–276.

Doria M. de F. (2010). Factors influencing public perception of drinking water quality. *Water Policy*, **12**(1), 1.

Dupont D. P. and Jahan N. (2012). Defensive spending on tap water substitutes: the value of reducing perceived health risks. *Journal of Water and Health*, **10**(1), 56.

Edwards M. (2004). Controlling corrosion in drinking water distribution systems: a grand challenge for the 21st century. *Water Science & Technology*, **49**(2), 1–8.

Edwards M., Powers K., Hidmi L. and Schock M. R. (2011). The role of pipe ageing in copper corrosion by-product release. *Water Science & Technology: Water Supply*, **1**(3), 25–32.

Engelhardt M., Savic D., Skipworth P., Cashman A., Saul A. and Walters G. (2003). Whole life costing: application to water distribution network. *Water Supply*, **3**(1–2), 87–93.

Espinosa-García A. C., Díaz-Ávalos R. and Solano-Ortiz R. (2014). Removal of bacteria, protozoa and viruses through a multiple-barrier household water disinfection system. *Journal of Water and Health*, **12**(1), 94–104.

Euzen A. (2003). How do consumers perceive water quality? Example of an anthropological study carried out in Paris. *Water Supply*, **3**(3), 263–269.

Fattahi P., Kherikhah A., Sadeghian R., Zandi S. and Fayyaz S. (2011). An evaluation model for measuring customer satisfaction levels in a water supply domain: case study of water supply in Hamedan. *Water Policy*, **13**(4), 490.

Fife-Schaw C., Barnett J., Chenoweth J., Morrison G. M. and Lundéhn C. (2008). Consumer trust and confidence: some recent ideas in the literature. *Water Science & Technology: Water Supply*, **8**(1), 43.

Franceys R. (2006). Customer committees, economic regulation and the Water Framework Directive. *Water Science & Technology: Water Supply*, **6**(5), 9.

Francisque A., Rodriguez M. J., Sadiq R., Miranda L. F. and Proulx F. (2011). Reconciling 'actual' risk with 'perceived' risk for distributed water quality: a QFD-based approach. *Journal of Water Supply: Research and Technology—Aqua*, **60**(6), 321.

Franquelo M. H. and Moises G. C. (2008). A social participation. Consumers and the water supply company. *Water Practice and Technology*, **3**(3), wpt2008070–wpt2008070.

Freeman M. C., Quick R. E., Abbott D. P., Ogutu P. and Rheingans R. (2009). Increasing equity of access to point-of-use water treatment products through social marketing and entrepreneurship: a case study in western Kenya. *Journal of Water and Health*, **7**(3), 527.

Freuze I., Brosillon S., Laplanche A., Tozza D. and Cavard J. (2005). Effect of chlorination on the formation of odorous disinfection by-products. *Water Research*, **39**(12), 2636–2642.

Gage G., Helmcke S. and Prigent O. (2013). The 21st century water utility initiative: how to boost operational performance of water utilities by up to 20%. *Water Science & Technology*, **68**(7), 1487.

Gallagher D. L. and Dietrich A. M. (2014). Statistical approaches for analyzing customer complaint data to assess aesthetic episodes in drinking water. *Journal of Water Supply: Research and Technology—Aqua*, **63**(5), 358.

Gebreegziabher K. and Tadesse T. (2011). Household demand for improved water supply services in Mekelle City, Northern Ethiopia. *Water Policy*, **13**(1), 125–142.

Gray J. (2008). Water contamination events in UK drinking-water supply systems. *Journal of Water and Health*, **6**(S1), s21–s26.

Groenendijk M., van de Wetering, S. and van Nieuwenhuijze R. (2008). Central water softening: customer comfort is relevant in new WHO view. *Water Science & Technology: Water Supply*, **8**(1), 69.

Hamouda M. A., Anderson W. B. and Huck P. M. (2012). Employing multi-criteria decision analysis to select sustainable point-of-use and point-of-entry water treatment systems. *Water Science & Technology: Water Supply*, **12**(5), 637–647.

Hanamoto T., Nagashio D. and Sasaki T. (2002). Integration of water treatment, environmental and information technologies: amagasaki project. *Water Supply*, **2**(5–6), 185–191.

Helland B. and Adamsson J. (1998). Performance indicators: benchmarking between six cities in Scandinavia. *Journal of Water Supply: Research and Technology—Aqua*, **47**(6), 284–288.

Hills S., Smith A., Hardy P. and Birks R. (2001). Water recycling at the Millennium Dome. *Water Science & Technology*, **43**(10), 287–294.

Hills S., Birks R. and McKenzie B. (2002). The Millennium Dome 'Watercycle' experiment: to evaluate water efficiency and customer perception at a recycling scheme for 6 million visitors. *Water Science & Technology*, **46**(6–7), 233–240.

Hoekstra E. J. (2008). Towards an improved human exposure assessment for contaminants in drinking water released by products used for domestic water distribution. *Water Practice & Technology*, **3**(1), wpt2008019.

Hofman J., Brink D., Breure P. and van der Hoek J. P. (2002). Fouling and accuracy drift of water meters. *Water Supply*, **2**(4), 129–136.

Hofman J. A. M. H., IJpelaar G. F., Heijman S. G. J., Vrouwenvelder J. S., Kruithof J. C. and Meer W. G. J. van der. (2005). Drinking water treatment in The Netherlands: outstanding and still ambitious. *Water Supply*, **4**(5–6), 253–262.

Hoogsteen K. J., Kolff G. van der and Ruijter J. A. (2008). Public private partnership for efficient and sustainable water supply development in Indonesia. *Water Practice & Technology*, **3**(1), wpt2008018.

Howard J. R., Lucas J., Maitland J., Tarrant P. and Watson T. (2005). WaterScope: an innovative water quality information management system. *Water Supply*, **4**(5–6), 409–414.

Howarth D. and Butler S. (2004). Communicating water conservation: how can the public be engaged? *Water Supply*, **4**(3), 33–44.

Huberts J. P. B., van Eekeren M. W. M. and Ernes S. (2000). New marketing tools for water companies. *Water Supply*, **18**(1), 493–496.

Hurlimann A. and Dolničar S. (2010). Acceptance of water alternatives in Australia. *Water Science & Technology*, **61**(8), 2137.

Husband P. S. and Boxall J. B. (2011). Asset deterioration and discolouration in water distribution systems. *Water Research*, **45**(1), 113–124.

Iacullo R. J. and Gradilone F. (2000). Customer service benchmarking: a market oriented approach to the delivery of water utility services. *Water Supply*, **18**(1), 500–503.

Ito T., Okumura T. and Yamamoto M. (1988). The relationship between concentration and sensory properties of 2-methylisoborneol and geosmin in drinking water. *Water Science & Technology*, **20**(8–9), 11–17.

Izaguirre G., Taylor W. D. and Pasek J. (1999). Off-flavor problems in two reservoirs, associated with planktonic pseudanabaena species. *Water Science and Technology*, **40**(6), 85–90.

Jaeger Y., Oberti S., Guichot L. and Baron J. (2006). Evaluation of treatment methods to reduce the corrosivity of soft waters. *Water Science & Technology: Water Supply*, **6**(5), 101–110.

Johnson K. (2002). Measuring up: do you have a scorecard? *Water Supply*, **2**(4), 181–187.

Johnson K. and Mortada M. M. (2005). Effective asset management maximizes your return on investment (ROI) and customer satisfaction. *Water Supply*, **5**(2), 97–106.

Johnson W. and MacCormick T. (2003). Issues of operational integrity in membrane drinking water plants. *Water Supply*, **3**(5), 73–80.

Jung S.-W., Baek K.-H. and Yu M.-J. (2004). Treatment of taste and odor material by oxidation and adsorption. *Water Science & Technology*, **49**(9), 289–295.

Juttner F. (1995). Physiology and biochemistry of odorous compounds from freshwater cyanobacteria and algae. *Water Science and Technology*, **31**(11), 69–78.

Kanakoudis V., Tsitsifli S., Samaras P. and Zouboulis A. (2013). Assessing the performance of urban water networks across the EU Mediterranean area: the paradox of high NRW levels and absence of respective reduction measures. *Water Science & Technology: Water Supply*, **13**(4), 939.

Karbowiak P. (2003). Marketing a technical business. *Water Supply*, **3**(1–2), 433–440.

Katko T. S. (1992). Cost recovery in water services. *Water Science & Technology*, **26**(9–11), 2621–2624.

Kavcar P., Odabasi M., Kiti M., Inal F. and Sofuoglu S. C. (2006). Occurrence, oral exposure and risk assessment of volatile organic compounds in drinking water for İzmir. *Water Research*, **40**(17), 3219–3230.

Kay D., Wyn-Jones A. P., Stapleton C. M., Fewtrell L., Wyer M. D., Watkins J. and Drury D. (2007). The microbiological quality of seven large commercial private water supplies in the United Kingdom. *Journal of Water and Health*, **5**(4), 523–538.

Kayaga S., Franceys R. and Sansom K. (2004). Bill payment behaviour in urban water services: empirical data from Uganda. *Journal of Water Supply: Research and Technology—Aqua*, **53**(5), 339–349.

Kitazawa H. (2006). Keeping residual chlorine and decreasing unpleasant odor caused by disinfection of tap water. *Water Supply*, **6**(2), 193–199.

Kjellberg S., Jayaratne A., Cadan E., Sukumaran N., Vreeburg J. and Verberk J. (2009). The resuspension potential method: Yarra Valley water's novel approach to routine mains cleaning. *Water Science & Technology: Water Supply*, **9**(5), 549.

Kumpel E. and Nelson K. L. (2013). Comparing microbial water quality in an intermittent and continuous piped water supply. *Water Research*, **47**(14), 5176–5188.

Levallois P., Grondin J. and Gingras S. (1999). Evaluation of consumer attitudes on taste and tap water alternatives in Quebec? *Water Science and Technology*, **40**(6), 135–140.

Liemberger R. (2009). Water Loss Assessment and Remediation in South and Southeast Asia. *Water Practice & Technology*, **4**(4), wpt2009077.

Lindhe A., Rosén L., Norberg T. and Bergstedt O. (2009). Fault tree analysis for integrated and probabilistic risk analysis of drinking water systems. *Water Research*, **43**(6), 1641–1653.

Liu B. and Reckhow D. A. (2015). Disparity in disinfection byproducts concentration between hot and cold tap water. *Water Research*, **70**, 196–204.

Loret J. F., Cossalter L., Robert S., Baudin I., Conan M. and Charles P. (2013). Assessment and management of health risks related to the recycling of filter backwash water in drinking water production. *Water Practice & Technology*, **8**(2), wpt.2013019.

Loureiro D., Alegre H., Coelho S. T., Martins A. and Mamade A. (2014a). A new approach to improve water loss control using smart metering data. *Water Science & Technology: Water Supply*, **14**(4), 618–625.

Loureiro D., Vieira P., Makropoulos C., Kossieris P., Ribeiro R., Barateiro J. and Katsiri E. (2014b). Smart metering use cases to increase water and energy efficiency in water supply systems. *Water Science & Technology: Water Supply*, **14**(5), 898–908.

Lubout K. L. and Lubout K. L. (2010). The role of water boards in water safety plan implementation in South Africa. *Water Practice & Technology*, **5**(1), 2.

Lukas A., Mayr E., Richard L. and Perfler R. (2011). Supporting the Water Safety Plan (WSP) approach with the Failure Experience Improvement System (FEIS). *Water Science & Technology: Water Supply*, **11**(3), 288–296.

Lytle D. A. and Snoeyink V. L. (2003). The effect of pH and dissolved inorganic carbon on the properties of iron colloidal suspensions. *Journal of Water Supply: Research and Technology—Aqua*, **52**(3), 165–180.

Lytle D. A., Sarin P. and Snoeyink V. L. (2005). The effect of chloride and orthophosphate on the release of iron from a cast iron pipe section. *Journal of Water Supply: Research and Technology—Aqua*, **54**(5), 267–281.

Mackey E. D., Baribeau H., Crozes G. F., Suffet I. H. and Piriou P. (2004). Public thresholds for chlorinous flavors in U.S. tap water. *Water Science & Technology*, **49**(9), 335–340.

Marchesan M. and Morran J. (2004). Tastes associated with products in contact with drinking water. *Water Science & Technology*, **49**(9), 227–231.

Martin R. J. and Shackleton R. C. (1990). Comparison of two partially activated carbon fabrics for the removal of chlorine and other impurities from water. *Water Research*, **24**(4), 477–484.

Martín-Alonso J., Devesa R., Bernal A. and Matia L. (2007). Managing an odour episode in Barcelona's water supply: strategies adopted, the causative agent (diacetyl) and determination of its organoleptic properties. *Water Science & Technology*, **55**(5), 209.

Matsui Y., Aizawa T., Kanda F., Nigorikawa N., Mima S. and Kawase Y. (2007a). Adsorptive removal of geosmin by ceramic membrane filtration with super-powdered activated carbon. *Journal of Water Supply: Research and Technology—Aqua*, **56**(6–7), 411–418.

Matsui Y., Aizawa T., Suzuki M. and Kawase Y. (2007b). Removal of geosmin and algae by ceramic membrane filtration with super-powdered activated carbon adsorption pretreatment. *Water Science & Technology: Water Supply*, **7**(5–6), 43–51.

May A. (2006). The benefits of drinking water quality regulation in England and Wales. *Water Science & Technology*, **54**(11–12), 387–393.

McDowall B., Hoefel D., Newcombe G., Saint C. P. and Ho L. (2009). Enhancing the biofiltration of geosmin by seeding sand filter columns with a consortium of geosmin-degrading bacteria. *Water Research*, **43**(2), 433–440.

McGuire M. J. (1995). Off-flavor as the consumer's measure of drinking water safety. *Water Science and Technology*, **31**(11), 1–8.

McGuire M. J., Loveland J., Means E. G. and Garvey J. (2007). Use of flavour profile and consumer panels to determine differences between local water supplies and desalinated seawater. *Water Science & Technology*, **55**(5), 275.

Mckenzie R. S. and van Rooyen P. G. van. (2003). Management of large water resource systems. *Water Supply*, **3**(3), 297–304.

Meij S. H. F. M., Ruiters C. J. M. and Stumphius C. J. (2005). Market-driven pricing structures for drinking water. *Water Supply*, **5**(6), 225–233.

Mendoza R. L. (2011). Fluoride-treated water and the problem of merit goods. *Water Policy*, **13**(1), 38.

Merkel W., Staben N., Brácker J., Levai P., Neskovic M., Petry D. and Weiss M. (2011). Structural conditions in German water supply. *Water Science & Technology: Water Supply*, **11**(5), 596.

Metaxas S. and Charalambous E. (2005). Residential price elasticity of demand for water. *Water Supply*, **5**(6), 183–188.

Miller M. (1993). The difficulties and promise of an inter-agency public information campaign for water quality issues. *Water Science & Technology*, **28**(3–5), 65–68.

Mons M. N., Heringa M. B., van Genderen J., Puijker L. M., Brand W. and van Leeuwen C. J. and van der Kooij D. (2013). Use of the Threshold of Toxicological Concern (TTC) approach for deriving target values for drinking water contaminants. *Water Research*, **47**(4), 1666–1678.

Morran J. and Marchesan M. (2004). Taste and odour testing: how valuable is training? *Water Science & Technology*, **49**(9), 69–74.

Morvay A. A., Decun M., Scurtu M., Sala C., Morar A. and Sarandan M. (2011). Biofilm formation on materials commonly used in household drinking water systems. *Water Science & Technology: Water Supply*, **11**(2), 252–257.

Mounce S., Machell J. and Boxall J. (2012). Water quality event detection and customer complaint clustering analysis in distribution systems. *Water Science & Technology: Water Supply*, **12**(5), 580.

Mounce S. R. and Boxall J. B. (2010). Implementation of an on-line artificial intelligence district meter area flow meter data analysis system for abnormality detection: a case study. *Water Science & Technology: Water Supply*, **10**(3), 437.

Mugabi J. and Njiru C. (2006). Marketing urban water services: information needs for water utility managers in developing countries. *Water Science & Technology: Water Supply*, **6**(5), 17.

Mugabi J., Kayaga S., Smout I. and Njiru C. (2010). Determinants of customer decisions to pay utility water bills promptly. *Water Policy*, **12**(2), 220.

Mugisha S. (2007). Performance assessment and monitoring of water infrastructure: an empirical case study of benchmarking in Uganda. *Water Policy*, **9**(5), 575–491.

Mullenger J., Ryan G. and Hearn J. (2002). A water authority's experience with HACCP. *Water Supply*, **2**(5–6), 149–155.

Muramoto S., Udagawa T. and Okamura T. (1995). Effective removal of musty odor in the Kanamachi purification plant. *Water Science and Technology*, **31**(11), 219–222.

Nema P., George K. V., Karemore M. G. and Kotangale J. P. (2004). Evaluation of water supply system infested with chironomus larvae. *Journal of Water Supply: Research and Technology—Aqua*, **53**(8), 573–580.

Neunteufel R., Perfler R., Mayr E., Theuretzbacher-Fritz H. and Kólbl J. (2010). Benefits from benchmarking: an Austrian case study. *Water Practice and Technology*, **5**(2), wpt2010046–wpt2010046.

Newcombe G. and Cook D. (2002). Influences on the removal of tastes and odours by PAC. *Journal of Water Supply: Research and Technology—Aqua*, **51**(8), 673–474.

Nicholson C. (2012). Delivering an essential and sustainable water plan for Sydney, Australia. *Water Science & Technology: Water Supply*, **12**(6), 747.

Nicholson N. F. (1993). Sustainability of water supplies in provincial Egypt. *Water Science & Technology*, **27**(9), 47–53.

Nyende-Byakika S., Ndambuki J. M. and Ngirane-Katashaya G. (2013). Modelling of pressurised water supply networks that may exhibit transient low pressure open channel flow conditions. *Water Practice & Technology*, **8**(3–4), wpt.2013054.

Ohkouchi Y., Yata Y., Bun R. and Itoh S. (2013). Chlorine requirement for biologically stable drinking water after nanofiltration. *Water Science & Technology: Water Supply*, **14**(3), 405–413.

Ómúr-Ózbek P. and Dietrich A. M. (2008). Developing hexanal as an odor reference standard for sensory analysis of drinking water. *Water Research*, **42**(10–11), 2598–2604.

Ömür-Özbek P., Little J. C. and Dietrich A. M. (2007). Ability of humans to smell geosmin, 2-MIB and nonadienal in indoor air when using contaminated drinking water. *Water Science & Technology*, **55**(5), 249.

Owen A. J., Colbourne J. S., Clayton C. R. I. and Fife-Schaw C. (1999). Risk communication of hazardous processes associated with drinking water quality: a mental models approach to customer perception, part 1: methodology. *Water Science and Technology*, **39**(10–11), 183–188.

Page B. (2003). Has widening participation in decision-making influenced water policy in the UK? *Water Policy*, **5**(4), 313–329.

Parena R. (2005). Deciding now what we want to be tomorrow as a key way of speeding the water business change. *Water Supply*, **5**(6), 53–56.

Parena R. and Smeets E. (2001). Benchmarking initiatives in the water industry. *Water Science & Technology*, **44**(2–3), 103–110.

Parinet J., Rodriguez M. J. and Sérodes J. (2010). Influence of water quality on the presence of off-flavour compounds (geosmin and 2-methylisoborneol). *Water Research*, **44**(20), 5847–5856.

Parker M. J., Casey R. J., Reynolds L. K., Vries R. R. de, Brueck T. M. and Williams C. N. (1993). Functional approach leads to successful ica across 124 water and 389 wastewater treatment works. *Water Science & Technology*, **28**(11–12), 37–43.

Pflieger G. (2006). The French model of water supply management challenged by users' empowerment. *Water Policy*, **8**(3), 211–229.

Piriou P. (2004). Chlorinous flavor perception in drinking water. *Water Science & Technology*, **49**(9), 321–328.

Piriou P., Devesa R., De Lalande M. and Glucina K. (2009). European reassessment of MIB and geosmin perception in drinking water. *Journal of Water Supply: Research and Technology—Aqua*, **58**(8), 532.

Polychronopolous M., Dudley K., Ryan G. and Hearn J. (2003). Investigation of factors contributing to dirty water events in reticulation systems and evaluation of flushing methods to remove deposited particles. *Water Supply*, **3**(1–2), 295–306.

Powe N. A., Garrod G. D., McMahon P. L. and Willis K. G. (2004). Assessing customer preferences for water supply options using mixed methodology choice experiments. *Water Policy*, **6**(5), 427–441.

Proulx F., Rodriguez M. J., Sérodes J. and Bouchard C. (2007). A methodology for identifying vulnerable locations to taste and odour problems in a drinking water system. *Water Science & Technology*, **55**(5), 177–183.

Proulx F., Rodriguez M. J., Sérodes J. B. and Miranda L. F. (2010). Factors influencing public perception and use of municipal drinking water. *Water Science & Technology: Water Supply*, **10**(3), 472–485.

Puffer M. and Al-Musallam L. B. A. (2007). Water stewardship in the desert: water conservation projects in the kingdom of Saudi Arabia. *Water Practice & Technology*, **2**(4), wpt2007074.

Puget S., Beno N., Chabanet C., Guichard E. and Thomas-Danguin T. (2010). Tap water consumers differ from non-consumers in chlorine flavor acceptability but not sensitivity. *Water Research*, **44**(3), 956–964.

Pybus P. and Schoeman G. (2001). Performance indicators in water and sanitation for developing areas. *Water Science & Technology*, **44**(6), 127–134.

Rachwal T. (2007). 30 Years of technical and organisational development in the UK water sector: Thames Water's experiences of moving from public to private sector. *Journal of Water Supply: Research and Technology—Aqua*, **56**(6–7), 419.

Rahman M. S. and Gagnon G. A. (2014). Bench-scale evaluation of drinking water treatment parameters on iron particles and water quality. *Water Research*, **48**, 137–147.

Ramaker A. and Pollemans L. C. (2008). Online metering changes the water industry. *Water Practice & Technology*, **3**(1), wpt2008013.

Rehan R., Knight M. A., Unger A. J. A. and Haas C. T. (2013). Development of a system dynamics model for financially sustainable management of municipal watermain networks. *Water Research*, **47**(20), 7184–7205.

Reid S., Chalak A. and Hecht J. (2010). Determining the optimal investment plan for water utilities: the case of Veolia Water Central. *Water Science & Technology: Water Supply*, **10**(3), 367.

Risebro H. L. and Hunter P. R. (2007). Surveillance of waterborne disease in European member states: a qualitative study. *Journal of Water and Health*, **5**(Suppl 1), S19.

Rivera V. R. (2011). The business of water: going the corporate way the case of Manila water. *Water Practice & Technology*, **6**(4), wpt20110079.

Rock C., Solop F. I. and Gerrity D. (2012). Survey of statewide public perceptions regarding water reuse in Arizona. *Journal of Water Supply: Research and Technology—Aqua*, **61**(8), 506.

Rogers D. M. (2005). Network analysis models: the secret for success. *Water Supply*, **4**(5–6), 383–388.

Rosén L., Lindhe A., Bergstedt O., Norberg T. and Pettersson T. J. R. (2010). Comparing risk-reduction measures to reach water safety targets using an integrated fault tree model. *Water Science & Technology: Water Supply*, **10**(3), 428.

Rousseau P. and Tranchant E. (2002). Business partners for development-KwaZulu-Natal pilot project, South Africa: a new approach to the delivery of sustainable water and sanitation services in poor peri-urban areas. *Water Supply*, **2**(4), 197–202.

Rouxel A., Brofferio S. and Guerin-Schneider L. (2008). Performance indicators and customer management: ACEA benchmarking experiences in water services in Latin America. *Journal of Water Supply: Research and Technology—Aqua*, **57**(4), 273.

Roy S., Phetxumphou K., Dietrich A. M., Estabrooks P. A., You W. and Davy B. M. (2015). An evaluation of the readability of drinking water quality reports: a national assessment. *Journal of Water and Health*, **13**(3), 645.

Ruas Dos Santos M. (2000). How is water perceived: as a commodity or as a product? *Water Supply*, **18**(1), 504–508.

Sakaki K., Matsuda Y., Hirayama N. and Itoh S. (2014). Development of comprehensive evaluation procedure for anti-seismic strategies: evaluating Kobe City's earthquake resistance improvement plan from the customer's viewpoint. *Journal of Water Supply: Research and Technology—Aqua*, **63**(2), 139.

Salman A., Al-Karablieh E. and Haddadin M. (2008). Limits of pricing policy in curtailing household water consumption under scarcity conditions. *Water Policy*, **10**(3), 295.

Sano H. (1988). The detection of taste and odor in Osaka's drinking water. *Water Science & Technology*, **20**(8–9), 37–42.

Sanz M. P., Schouten M. and Hantke-Domas M. (2011). Consumer-inclusive economic regulation: water pricing under different regulatory regimes. *Water Policy*, **13**(4), 459–470.

Satchwill T., Watson S. B. and Dixon E. (2007). Odourous algal-derived alkenes: differences in stability and treatment responses in drinking water. *Water Science & Technology*, **55**(5), 95.

Seth A., Bachmann R., Boxall J., Saul A. and Edyvean R. (2004). Characterisation of materials causing discolouration in potable water systems. *Water Science & Technology*, **49**(2), 27–32.

Seyoum A. G., Tanyimboh T. T. and Siew C. (2013). Assessment of water quality modelling capabilities of EPANET multiple species and pressure-dependent extension models. *Water Science & Technology: Water Supply*, **13**(4), 161–116.

Shafiee M. E. and Zechman E. M. (2013). An agent-based modeling framework for sociotechnical simulation of water distribution contamination events. *Journal of Hydroinformatics*, **15**(3), 862–880.

Sly L. I., Hodgkinson M. C. and Arunpairojana V. (1989). The importance of high aesthetic quality potable water in tourist and recreational areas. *Water Science & Technology*, **21**(2), 183–187.

Speers A., Burn S., MacDonald D. H., Young M. and Syme G. (2003). Setting and evaluating customer service standards. *Water Supply*, **3**(1–2), 381–388.

Stine S. W., Pepper I. L. and Gerba C. P. (2005). Contribution of drinking water to the weekly intake of heterotrophic bacteria from diet in the United States. *Water Research*, **39**(1), 257–263.

Suffet I. H., Burlingame G. A. and Mackey E. (2008). Controlling taste and odour events for the city of Philadelphia, USA: a case study. *Water Science & Technology: Water Supply*, **8**(2), 135–141.

Terpstra P. M. J. (1999). Sustainable water usage systems: models for the sustainable utilization of domestic water in urban areas. *Water Science and Technology*, **39**(5), 65–72.

Tondelier C., Benanou D., Thouvenot T. and Genin A. (2008). Novel on-tap passive sampling for odorous compounds in drinking water. *Water Science & Technology: Water Supply*, **8**(2), 197–205.

Trietsch E. A. and Vreeburg J. H. G. (2005). Reliability of valves and section isolation. *Water Supply*, **5**(2), 47–51.

Uhl W., Schaule G. and Gimbel R. (2002). Preventing bacterial regrowth in old distribution systems without disinfection. *Water Supply*, **2**(3), 259–266.

van de Veerdonk, J., Van Essen C. and Dekker J. (2008). Risks and keeping customers' confidence: the key role of consumers' perception regarding the safety of tap water. *Water Practice & Technology*, **3**(1), wpt2008017.

Van Houtte E. and Verbauwhede J. (2008). Torreele's water re-use facility enabled sustainable groundwater management in de Flemish dunes (Belgium). *Water Practice & Technology*, **3**(2), wpt2008039.

Van Lieverloo J. H. M., Bosboom D. W., Bakker G. L., Brouwer A. J., Voogt R. and Roos J. E. M. D. (2004). Sampling and quantifying invertebrates from drinking water distribution mains. *Water Research*, **38**(5), 1101–1112.

Van Lieverloo J. H. M., Hoogenboezem W., Veenendaal G. and Kooij D. van der. (2012). Variability of invertebrate abundance in drinking water distribution systems in the Netherlands in relation to biostability and sediment volumes. *Water Research*, **46**(16), 4918–4932.

van Thienen P. (2013). A method for quantitative discrimination in flow pattern evolution of water distribution supply areas with interpretation in terms of demand and leakage. *Journal of Hydroinformatics*, **15**(1), 86–102.

Von Seidlitz A. E. and Londong J. (2001). Modern management instruments—control of a water resources management association by means of an agreement on targets (balanced scorecard). *Water Science & Technology*, **44**(2–3), 119–126.

Vreeburg J. H. G. and Boxall J. B. (2007). Discolouration in potable water distribution systems: a review. *Water Research*, **41**(3), 519–529.

Vreeburg J. H. G., Schippers D., Verberk J. Q. J. C. and Van Dijk J. C. (2008). Impact of particles on sediment accumulation in a drinking water distribution system. *Water Research*, **42**(16), 4233–4242.

Webber M. A., Atherton P. and Newcombe G. (2015). Taste and odour and public perceptions: what do our customers really think about their drinking water? *Journal of Water Supply: Research and Technology—Aqua*, **64**(7), 802–811.

Whelton A. J. and Dietrich A. M. (2004). Relationship between intensity, concentration, and temperature for drinking water odorants. *Water Research*, **38**(6), 1604–1614.

Whelton A. J., Dietrich A. M., Burlingame G. A., Schechs M. and Duncan S. E. (2007). Minerals in drinking water: impacts on taste and importance to consumer health. *Water Science & Technology*, **55**(5), 283–291.

White I. (2010). Rainwater harvesting: theorising and modelling issues that influence household adoption. *Water Science & Technology*, **62**(2), 370.

Whittington D. (2003). Municipal water pricing and tariff design: a reform agenda for South Asia. *Water Policy*, **5**(1), 61–76.

Wiesenthal K. E., McGuire M. J. and Suffet I. H. (2007). Characteristics of salt taste and free chlorine or chloramine in drinking water. *Water Science & Technology*, **55**(5), 293–300.

Wijkman A. and Gordon L. (2002). Workshop 9 (synthesis): how to increase the status of water issues in governance and in public perception. *Water Science & Technology*, **45**(8), 229–231.

Wyatt A. and Alshafey M. (2012). Non-revenue water: financial model for optimal management in developing countries – application in Aqaba, Jordan. *Water Science & Technology: Water Supply*, **12**(4), 451.

Xie P., Ma J., Liu W., Zou J., Yue S., Li X. and Fang J. (2015). Removal of 2-MIB and geosmin using UV/persulfate: contributions of hydroxyl and sulfate radicals. *Water Research*, **69**(1), 223–233.

Yates D. N., Lavinb F. V., Purkeyc D. P., Guerrerod S., Hanemanne M. and Sieber J. (2013). Using economic and other performance measures to evaluate a municipal drought plan. *Water Policy*, **15**(4), 648–668.

Young W. F., Horth H., Crane R., Ogden T. and Arnott M. (1996). Taste and odour threshold concentrations of potential potable water contaminants. *Water Research*, **30**(2), 331–340.

Zander A. K. and Pingert P. (1997). Membrane-based extraction for detection of tastes and odors in water. *Water Research*, **31**(2), 301–309.

Zubizarreta J. (2008). Drought management and customer service. *Water Practice & Technology*, **3**(1), wpt2008015.

IWA Publishing's authorised EU representative for General Product Safety Regulations is Diane D'Arras, 15 rue Duret, 75116 Paris, France, e-mail: safety@iwap.co.uk.

Printed and bound by CPI Group (UK) Ltd, Croydon, CR0 4YY
12/05/2026
02108891-0003